USMLE

SERIES

CONCEPTS IN GROSS ANATOMY
A review for the USMLE Step 1

USMLE
Concepts
SERIES

CONCEPTS IN GROSS ANATOMY
A review for the USMLE Step 1

by

William T. Mosenthal, MD

Department of Anatomy
Dartmouth Medical School
Hanover, New Hampshire, USA

The Parthenon Publishing Group
International Publishers in Medicine, Science & Technology

NEW YORK   LONDON

Published in the USA by
The Parthenon Publishing Group Inc.
One Blue Hill Plaza, Pearl River, New York 10965, USA

Published in the UK and Europe by
The Parthenon Publishing Group Ltd.
Casterton Hall, Carnforth, Lancs. LA6 2LA, UK

Library of Congress Cataloging-in-Publication Data
Mosenthal, William T.
    Concepts in gross anatomy : a review for the USMLE step 1 / by William
T. Mosenthal.
        p.   cm.   – –   (USMLE concepts series)
        ISBN 1–85070–928–9
        1. Human anatomy.   2. Human anatomy – – Examinations, questions, etc.
    I. Title.   II. Series
    QM23.2.M683   1997
    611'.0076– –dc21   96-47926
                                                                              CIP

British Library Cataloguing in Publication Data
Mosenthal, William T.
    Concepts in gross anatomy : a review for the USMLE step 1. -
    (USMLE concepts series)
    1. Human anatomy - Examinations - Study guides
    I. Title
    611'.0076

ISBN 1-85070-928-9

Typeset by H&H Graphics, Blackburn, UK
Printed and bound in the USA

Contents

Preface

A competent clinician is a good anatomist. A good clinician uses this truly basic knowledge daily to understand the signs, symptoms, progress and treatment of the disorders encountered. In this book, I have attempted to describe the anatomy that I have found important in my own clinical experience, excluding trivial and clinically irrelevant material. Although no two people have ever fully agreed as to the appropriate content of an anatomical 'core', I am hopeful that this volume will prove useful in the identification, emphasis and clarification of clinically important anatomical material.

William T. Mosenthal, MD

Professor of Anatomy and
Clinical Surgery, Active Emeritus
Dartmouth Medical School
Hanover, New Hampshire

1 Back

The anatomy of the back is chiefly concerned with the vertebral column. The vertebral column supports the weight of the body in the human upright posture, allowing movement of the trunk and neck while protecting the spinal cord.

Vertebrae.   All have a common basic arrangement: body, pedicles, laminae, spine, transverse processes, intervertebral facet joints (superior and inferior pairs), vertebral notches, spinal canal and intervertebral discs.

The **body**, a round block of bone, supports the weight. The **pedicles** and **laminae** meet to form a bony **spinal canal** in which the spinal cord resides. Where the laminae and pedicles meet, a **transverse process** protrudes laterally. Where the two laminae meet, a **spinous process** protrudes in the midline posteriorly. These projections offer points of attachment for spinal muscles and supporting ligaments. Superior and inferior pairs of **articular facets** protrude upward and downward from the pedicle-lamina junctional area, permitting flexion, extension, lateral flexion and twisting of the spine. The **vertebral notches** are cut out of the superior and inferior aspects of the bases of the pedicles. When the vertebrae are fitted together, the notches of adjoining vertebrae meet to form the **intervertebral foramina** through which the spinal nerves exit. The dorsal root ganglion resides in the intervertebral foramen.

Between the bodies lies the shock-absorbing **intervertebral disc** which consists of a central compressible **nucleus pulposus** held together by a surrounding **annulus fibrosis**. The nucleus pulposus articulates with the hyaline cartilage covering the superior and inferior surfaces of the vertebral bodies. The nucleus pulposus is the adult remnant of the embryonic notochord.

Spinal column.   The spinal column consists of:

7	cervical vertebrae
12	thoracic vertebrae
5	lumbar vertebrae
5	sacral (fused) vertebrae
3–5	coccygeal (fused) vertebrae.

Vertebral bodies are banded together by **anterior** and **posterior longitudinal ligaments**. Strong supporting ligaments run between vertebral spines and between transverse processes. Vertebral laminae are bound together, thus closing the spinal canal, by a yellowish elastic **ligamentum flavum** (L. yellow).

All vertebrae have the same fundamental structure, but each area has its individual characteristic features. Cervical vertebrae demonstrate foramina in the tips of their

1

transverse processes for the passage of the vertebral artery. The thoracic vertebrae have the joint facets for the ribs on the body and transverse processes. Lumbar vertebrae are big and sturdy with large blunt straight spines. The important atlas and axis – C1 and C2, respectively – are highly specialized and described in detail in the Neck section.

The sacrum is composed of five vertebrae fused to form a single bone. It transfers the body weight to the pelvic girdle through the sacroiliac joints. These are, in fact, synovial joints but permit little motion. They are bound firmly by strong sacroiliac ligaments. Posteriorly, there is an open gutter, the **sacral hiatus**, due to the absence of laminae and spinous processes on S4–5. The distal ends of the hiatus are marked on each side by enlargements (**sacral cornua**) which are readily palpable at the superior extremity of the intergluteal cleft.

The **coccyx** consists of vestigial bones and nerves of the vestige of the tail. This is a functionally unimportant part of the column, but productive of pain (coccygodynia) if traumatized.

The **spinal canal** houses the spinal cord, which is encased in **meninges** (Gk. membranes) and floats in **cerebrospinal fluid** (CSF). The membranes consist of an outer tough **dura** loosely attached to the bony canal, a middle **arachnoid** and an inner intimate **pia**, the latter snugly attached to the spinal cord. Outside of the dura is a voluminous extradural venous plexus. There is a gross space between the pia and arachnoid termed the **subarachnoid space**. This space contains the CSF, which is formed in the brain, and floats the entire central nervous system (CNS). The arachnoid and dura line the bony spinal canal with only a potential space between them. The subarachnoid space ends at the S2 level; the spinal cord itself ends at the L2 level. Thus, there is a cord-free cistern (**lumbar cistern**) lying between the ends of the cord and the subarachnoid space. A needle may be inserted through the interspinous ligament and dura–arachnoid between the spines of L3–4 or L4–5 to enter this cistern. A sample of CSF may thereby be obtained without endangering the neurons of the cord. The space is traversed by nerve fibers from the terminal cord destined to make up spinal nerves L3–S5, descending through the cistern to the level of the appropriate foramina. This bundle of nerve fibers is known as the **cauda equina** (L. horsetail). These fibers will not be damaged by a needleprick. 'Denticulate' pial ligaments attach to the dura and protect the spinal cord from severe back-and-forth lateral motion. A similar tethering effect is afforded by the **pial filum terminale** which extends from the end of the spinal cord to the dura at the caudal end of the lumbar cistern. CSF is important in determining CSF pressure and flow, and in the diagnosis of inflammation or tumor, for example, in the CNS. Spinal anesthesia may be induced by injecting the anesthetic agent into the lumbar cistern.

The sacral hiatus is also easily entered with a needle after identifying the sacral cornua. Here the needle is extradural (dural envelope ceases at S2). Any anesthetic injected here will affect the most distal nerve fibers (S3–5) and offers a means of anesthetizing the anal canal and perianal area (caudal block).

The spinal column is not straight up and down. There are four normal curves: the cervical and lumbar curves are convex anteriorly; the thoracic and sacral curves are concave anteriorly. Lateral curves are not normal (scoliosis). Exaggerated lumbar and cervical curves are termed 'lordosis' whereas exaggerated thoracic curvature is called 'kyphosis'. The normal weight-bearing line in the erect position passes just anterior to the sacral promontory.

Pertinent surface anatomy permits identification of vertebrae in the column: **vertebra prominens** (C7 spine, the first spine palpable in the flexed neck); the prominent spines of T1–3; the 'shingled' spines of T4–12 (due to the overlapping of these spines, the palpable spine tip lies over the body of the next vertebra down; for example, the spine of T6 lies over the body of T7); and the palpable straight-back lumbar spines.

It is important that an understanding of this anatomy be kept firmly in mind. Disorders of the back are legion. Areas of greatest stress are those where motion and curvature are greatest, the cervical and lumbar areas. Hyperflexion and hyperextension give rise to tearing of the anterior and posterior ligaments and musculature with resulting pain ('whiplash'). Damage or degeneration of the annulus fibrosus permits escape of the nucleus pulposus. The protrusion of the nucleus beyond its normal confines results in pressure on the spinal nerves exiting through the vertebral foramina – the familiar 'ruptured disc'. Arthritic or traumatized facet joints result in painful movement and may also press on neighboring nerves. And so on! The highly specialized vertebrae and joints of C1 and C2 (atlas and axis) are described in the Neck section.

Muscles of the back. These include the muscles that move the spine: the deep muscles; and those that run from the spine to the pectoral girdle – the superficial muscles.

Deep back muscles. These consist of a deep group and a more superficial group. The deepest group consists of a bewildering array of small and large muscles occupying the space between the transverse processes and the spines. These are multifidus, semispinalis, rotatores, interspinales, intertransversarii, and the levators costarum. Just know that they are there! They help extend, flex, twist and lateral flex. The more superficial muscles of the deep muscles may be lumped together and called **sacrospinalis** or **erector spinae**. This massive muscle group does just what its name suggests it should - erects (extends) the spine. It may be divided into three poorly defined segments, all arising from a huge tendon attached to the posterior iliac crest, posterior sacrum, and lumbar spines and supraspinous ligaments. The medial portion is spinalis, which inserts onto the thoracic spines. The intermediate column is longissimus, which inserts into the transverse processes of all higher vertebrae. The lateral column is iliocostalis, which inserts into the angles of the ribs. As these muscle masses are not entirely discrete, it is well to consider the whole simply as erector spinae.

This entire mass of deep musculature is innervated by the **dorsal** primary rami of all the spinal nerves down to the sacrals.

Superficial back muscles. Of least significance are the small superior and inferior serratus posterior muscle pairs that lift the ribs and resist the inward pull of the diaphragm on the ribs, respectively. Innervation is via the dorsal primary rami.

Of great importance are the muscles running from the spinal column across the back to the pectoral girdle. These include levator scapulae, the rhomboids, latissimus dorsi and trapezius. All of these are important in the function of the upper extremity and are described in the section on the Upper Extremity. All are innervated by the **ventral** primary rami of the brachial plexus except for trapezius, which is innervated by the spinal accessory nerve.

Back Self-Assessment

Identify lettered items.

1. Lamina.
2. Pedicle.
3. Superior articular facet.
4. Facet for articulation with rib tubercle.
5. Transverse process.
6. Body.
7. Spinous process.
8. Facet for articulation with head of rib.

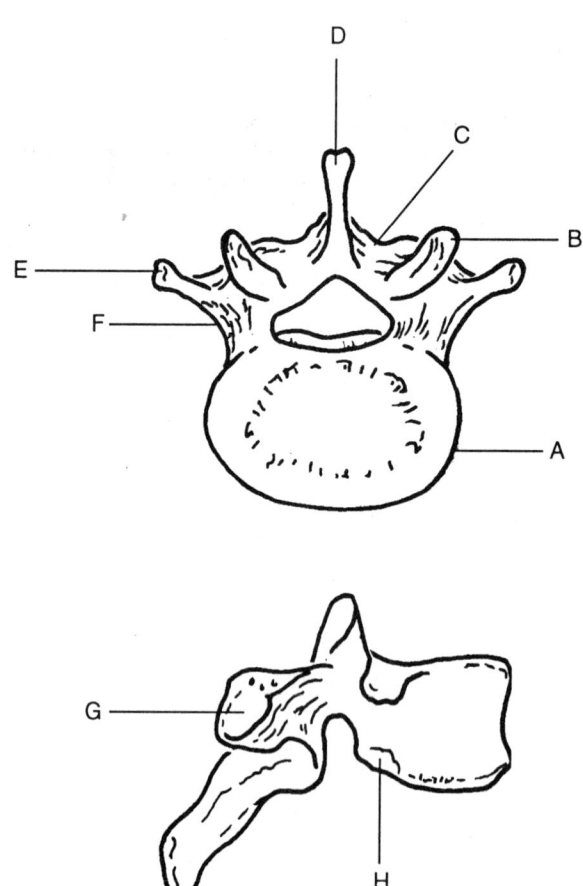

Match each lettered item with a numbered item:

    A. Cervical vertebra
    B. Thoracic vertebra
    C. Lumbar vertebra
    D. Sacrum

9. Vertebra prominens.

10. Largest single vertebra.

11. Vertebral transverse process contains a vascular foramen.

12. Vertebal canal is open posteriorly.

13. Sampling of cerebrospinal fluid is done in the lower lumbar region because:

    A   The caudal extremity of the spinal cord is at L2.
    B   The subarachnoid space extends to L2.
    C   The cerebrospinal fluid is less agitated and thus clearer at lumbar levels.
    D   There is no vertebral venous plexus at level L4.
    E   The low lumbar cord is largely vestigial and, thus, if injured, little damage is done.

14. Which statement is incorrect?

    A   A thoracic vertebra may contain twelve joints.
    B   The posterior longitudinal ligament lies between the spinous processes.
    C   The intervertebral disc allows some movement between vertebral bodies.
    D   The ligamentum flavum extends between the vertebral laminae.
    E   All back muscles except those extending to the pectoral girdle or upper extremity proper are innervated by dorsal primary rami of the spinal nerves.

Answers

1. **C**
2. **F**
3. **B**
4. **G**
5. **E**
6. **A**
7. **D**
8. **H**
9. **A**
10. **C**
11. **A**
12. **D**
13. **A**
14. **B**

② Thorax

The thorax is the chest, the torso between the neck and the abdomen, and is formed by 12 thoracic vertebrae, 12 pairs of attached ribs and the sternum. It is separated from the abdomen by a muscular diaphragm, and from the neck by muscles and fascia running from the first and second ribs to the cervical spine.

The thoracic cavity is a closed space contained by a jointed cage of movable ribs which are attached anteriorly to the sternum and posteriorly to the spinal column. The cavity is closed at the top by the fascia and muscle running from ribs to neck and, at the bottom, by the diaphragm. It is divided in the middle by a septum, the mediastinum, producing two cavities. Each of these is lined by an airtight slippery membrane, the pleura. The mediastinum contains the heart, enclosed in its own slippery pericardial sac, and the respiratory airway, which branches and expands out into the lungs which fill each pleural cavity. Passing through the mediastinum is the gut tube, the esophagus. Arteries, veins, lymphatics and nerves passing to and from the cardiorespiratory organs traverse the superior mediastinum. Penetrating the diaphragm at the inferior base are the vessels and nerves servicing the lower body structures. There are three large openings in the diaphragmatic dome for these: aortic hiatus at T12; esophageal hiatus at T10; and hiatus for the inferior vena cava at T8.

To follow is a review of the clinically significant anatomy of all the above.

Thorax wall: Skin and breast.   The muscles running from ribs to neck and pectoral girdle are studied with the Upper Extremity and Neck regions. Many of these, usually considered in reference to their action with the ribs 'fixed', can also work the other way. 'Fixation' may be in their girdle–neck attachments, and their contraction then raises the ribs. Such muscles (scalenes, sternocleidomastoid and pectoralis minor, among others) constitute auxiliary muscles of respiration.

Skin.   On sectioned skin, three distinct strata are identified: the tough epidermis-dermis; the fatty superficial layer of subcutaneous fascia; and the fibrous membranous deep layer of the subcutaneous fascia. All skin is movable over the true deep fascia that covers all muscle. Note the areolar layer between the deep muscle fascia and deep layer of superficial fascia of the skin of the thorax that permits this motion. Variations in thickness and strength of these layers will become obvious as dissection progresses. In many areas, the deep layer of superficial fascia is important in the suture of skin wounds. (Note that 'subcutaneous' and 'superficial' are used interchangeably.)

Breast.   This is a skin gland. It is embryologically related to the sweat and sebaceous glands, and all are embryological ingrowths of epidermal epithelium.

Nipple.   Contains the excretory ducts (lactiferous ducts) of the breast glands. Minute duct openings may be seen on the apical nipple surface.

Areola.   Pigmented area around the nipple, roughened by the presence of special lubricating glands important in nursing (Montgomery's glands).

Internal structure.   The size of the breast is largely due to its fat content (and functional state). Breast glands are arranged in 20 or more lobes, demarcated and suspended by connective tissue strands extending from the deep layer of the subcutaneous fascia to the epidermis. These are called the suspensory ligaments of the breast or Cooper's ligaments. Glandular tissue is firm and white in contrast to the soft yellow fat in which the lobes reside. Each lobe is served by an excretory duct which exits through the nipple.

Not much of the above-described can be dissected out in the senile breasts of most female cadavers. These structures are described here because each element frequently causes clinical problems. Obstructed areolar Montgomery's glands or lactiferous ducts are common causes of breast infection. Breast glands and ducts are often sites of cancer. Cooper's ligaments may contract in the usual desmoplastic reaction to cancer and produce a distortion of the overlying breast skin to which they are attached. Such cancer-caused skin dimpling may occur even when no tumor can be palpated. Cancers that grow deeply through the deep layer of the subcutaneous fascia may involve the deep muscle fascia and inhibit the normal free motion of the breast over the chest wall. The nipples may drain various kinds of fluid, depending on the type of glandular or ductal pathology present (for example, pus, blood, clear fluid or milk).

Knowledge of the breast skin anatomy allows an understanding of these signs.

Multiple draining ducts indicate a general process. A solitary draining duct indicates an isolated pathology. As the ducts are arranged like the spokes of a bicycle wheel, gentle pressing (using the eraser end of a pencil) will produce drainage when applied over the involved area. This anatomical identification permits accurate local investigation.

Male breast.   A vestigial organ, the male breast still, however, contains ducts. Male breast enlargement (gynecomastia) may occur due to estrogen medication taken for cancer of the prostate, in liver disease in which the liver is unable to destroy the small amounts of estrogen normally manufactured by the male, at birth and at puberty because of hormonal imbalances and, rarely, due to cancer of the male breast ducts.

Milk line.   This is a ridge in the human embryo that runs from axilla to inguinal region. In animals, multiple breasts are formed along this line. In humans, normally only the two pectoral breasts develop. However, a nipple, a nipple and areola and, occasionally, an entire breast may form anywhere along this milk line. Ectopic or supernumerary nipples are common. Axillary 'tumors' are frequently normal axillary breast tissue.

Lymph drainage.   Lymphatics of the breast cannot be dissected; they are too tiny. Cancer of the breast can and does spread through breast lymphatics to regional nodal areas, particularly the ipsilateral axillary nodes along the axillary vein and the ipsilateral nodes along the anterior thoracic vessels. Thus, these areas must be carefully examined when evaluating a breast lump.

Bones and joints of the thorax.   Now is the time to review the vertebral components of the spine. They all have a common basic arrangement and include a body, pedicle, lamina, spine, transverse process, intervertebral joint facets (superior and inferior pairs), intervertebral notch, spinal canal and intervertebral disk. (Thoracic vertebrae also have articular facets for ribs.)

The body supports weight. The pedicles (little feet) and the laminae meet to form a bony canal in which the spinal cord resides. Where the laminae and pedicles meet, a transverse process extends laterally. Where the two laminae meet, a spinous process protrudes posteriorly. These projections offer points of attachment for spinal muscles and supporting ligaments. Superior and inferior pairs of articular facets protrude upward and downward from the pedicle-lamina junctional area, permitting flexion, extension and twisting spine motion. The intervertebral notches, transmitting the segmental spinal nerves, are cut out of the superior and inferior aspects of the pedicles.

Lying between the vertebral bodies is the shock-absorbing intervertebral disk, which consists of a central compressible nucleus pulposus and a surrounding annulus fibrosis. There is a considerable space between the pedicles of adjacent vertebrae – the intervertebral foramen formed where adjacent pedicular intervertebral notches meet, through which spinal nerves pass to and from the spinal cord. It is easy to see that any extrusion of the gelatinous nucleus pulposus through a defective annulus fibrosus could press on a spinal nerve traversing the intervertebral foramen and could produce pain and paralysis in the body segment supplied by that nerve.

These terms and this anatomy should be firmly fixed in your mind. You should be able to sketch a typical vertebra. You should then be able to understand the symptoms produced by injuries such as ruptured discs or fractured spines, or facet-joint problems – common problems all! Spine films will become intelligible.

Ribs.   The typical rib possesses a head, neck, tubercle, angle, shaft, costal groove and costal cartilage. The rib angle is lateral to the tubercle, identified by a ridge for attachment of the iliocostalis muscles. The tubercle is the site of articulation with the vertebral transverse process and the attachment of ligaments. Ribs are jointed with elements of the thoracic vertebrae and the sternum; they extend forward and downward from their vertebral attachments, and are capable of motion at their joints in two directions. The rib can move outwards like the lifting of a bucket handle. In this motion, the rib rotates outwards around a horizontal axis perpendicular to the neck of the rib. This movement results in an increase in the side-to-side diameter of the thorax. The rib can also move upwards around its vertebral attachment and the rib neck rotates around its own axis. This movement is like lifting a pump handle; the sternum rises and the anteroposterior diameter of the thorax is increased.

The jointing of the ribs and vertebrae is complex. Typically, the head is jointed to the demifacets of adjacent vertebrae and the rib tubercle is jointed to the tip of the vertebral transverse process. It is important only to know that these joints exist and are at work with every breath. There are differences here and there which are trivial at this stage. A good hard look at the skeleton should fix the general joint appearances in your mind.

Ribs are moved by intercostal muscles and 'external muscles of respiration'. The latter include particularly the sternocleidomastoid, scalenes, pectoralis minor and serratus anterior. These lift the ribs and sternum when the skull, spine or pectoral (shoulder) girdle serve as fixed origins. The exact mode of action of the intercostal muscles is uncertain. It is unnecessary at this time to define all the parts of the intercostal musculature. They are there and do an important job; their exact structure and nomenclature are not essential. These muscles lie in three levels: external; internal; and innermost (*cf* abdominal wall muscles in the Abdomen section). Intercostal vessels and nerves nourish and innervate these muscles and the other tissues of each intercostal segment. They run partially protected in the costal grooves in the inferior border of each rib, with the vein most cranial and the nerve most caudal. They give off lateral and anterior branches in their course to supply the needs of the more superficial parts of the wall segment.

The anterior attachments of the ribs, to complete the thoracic cage, are either to the sternum by means of synovial sternocostal joints (ribs 1 through 6) or to each other by joining costal cartilages to form a common cartilaginous costal margin ending in a single costosternal synovial joint at the caudal sternum (ribs 7 through 10). The eleventh and twelfth ribs are free at their tips.

'Slipped ribs' are tears through the common cartilage of the costal margin or through the costosternal joints. This is a common injury especially for people such as quarterbacks who get hit while passing, which leaves the rib cage unprotected by the arm. This common debilitating injury has led to the wearing of stiff padded 'flak-jackets' by these players.

Sternum. Make an effort to understand the three segments of the sternum – manubrium, body and xiphoid process – and the anatomical importance of the sternal angle, the angle of Louis (a 19th-century French physician.). The angle between the manubrium and body is easily palpated and always accepts the end of the second rib, making rib-counting easy. The angle marks the level of bifurcation of the trachea, the beginning and end of the aortic arch, the junction between the T4 and T5 vertebrae, and is the front edge of the horizontal plane that defines the boundary between the superior and inferior mediastinal areas.

The sternum is clinically important as a site for bone marrow aspiration. It is a frequent victim of steering-wheel blows in auto accidents, which often results in injury to the mediastinal structures it is designed to protect – the heart and great vessels of the superior mediastinum.

Spinal nerves. We are fundamentally segmented structures. Each segment has its nerve. The thoracic cage is a good example of this segmental construction and is a good place to demonstrate a segmental nerve. Each spinal nerve is a mixed nerve with motor, sensory and autonomic content. There are 12 thoracic segments and, thus, 12 thoracic spinal nerves; each spinal nerve issues from an intervertebral foramen and is numbered according to the number of the vertebra whose inferior pedicle notch forms the upper border of the foramen; for example, the nerve issuing from under the inferior intervertebral notch of T6 is spinal nerve T6.

Note that the notch, and thus the T6 foramen, is caudal to the pedicle of that vertebra.

Somatic components of the nerve supply both motor and sensory elements of the segment. The autonomic portion is more complex. Blood vessel smooth muscle, and segmental sweat and sebaceous glands require autonomic postganglionic nerve fibers. Emerging in the spinal nerve are preganglionic fibers originating in spinal cord neurons. Soon after egress, there is a branch that runs from the spinal nerve to the sympathetic trunk. The trunk is a long ganglionated chain running downwards subpleurally across the region of the necks of the ribs and will be defined later when we discuss the posterior mediastinum. A second branch runs from the trunk back to the spinal nerve. This second branch contains the necessary postganglionic fibers which have arisen from neurons in the sympathetic chain ganglion. The first branch is called a white ramus communicans, the second a gray ramus communicans. 'White' and 'gray' refer to the presence or virtual absence of myelin in these nerves. Every spinal nerve has a gray ramus running to it from the sympathetic trunk and all thoracic spinal nerves also have a white ramus branch. However, as white ramus preganglionic neurons are only found from T1–L2 levels of the spinal cord, only T1–L2 spinal nerves emit white rami. The others receive their preganglionic fibers from white rami originally given off by spinal nerves T1–L2, but which have had to travel up or down in the sympathetic trunk to the appropriate matching sympathetic trunk ganglion for synapse and emergence as gray rami. The sympathetic trunk runs along the entire extent of the vertebral column – from the base of the skull to the coccyx – and thus can service all spinal nerves and all body segments.

Pleural cavity. This is the space inside the thoracic cage, lined with the pleural membrane and containing the lungs. Its lower extent is the respiratory diaphragm and its upper limit is the cupola of the pleura. The cupola extends several centimeters up into the neck above the level of the clavicle. Look at a chest x-ray; pass your hand up into the cupola of your cadaver to verify this. The base of the pleura covers the dorsal surface of the diaphragm. At the periphery of the cavity, the pleura is attached to the chest wall at the level of the attachments of the diaphragm. As the diaphragm bulges up into the center of each pleural cavity, the center of the cavity is considerably higher than the periphery. It is imperative that you know the surface anatomy of the periphery and dome-like center of the pleural cavity base. Start with the tip of rib 6 at the xiphisternal junction, the anterior midpoint attachment level of the diaphragm. The inferior periphery of the cavity – the diaphragmatic attachment – extends laterally to cross rib 8 in the vertical midclavicular line and rib 10 at the vertical midaxillary line, and follows rib 12 from the scapular line (vertical line of the medial border of the scapula) to the twelfth vertebra. The dome of the diaphragm reaches as high as the fourth interspace in the midclavicular line. The space between the upward bulging diaphragm and the chest wall is called the **costophrenic recess.** Tucked under the thin diaphragm is the peritoneal cavity – only millimeters away from the pleural cavity – with the liver on the right, and the stomach fundus, spleen and splenic colon flexure on the left pushing upwards next to the heart and lungs.

Accurate knowledge of pleural cavity surface anatomy enables assessment of the shadows seen on x-rays, possible damage done by penetrating wounds to the neck, chest or abdomen, possible source of sounds heard by stethoscope, etc. Spill of bile, stomach acid, blood or feces into the pleural cavity can be disastrous and must be promptly recognized. The two pleural cavities are separated by the midline mediastinum. The right and left pleurae practically touch each other between the distal esophagus and the descending aorta. This is important anatomy for the chest surgeon who seeks to keep one pleural cavity intact.

Lungs. The right and left lungs bulge into and practically fill the pleural cavities. By pushing their way into the pleural cavities in their development, they each receive a coating of pleural membrane - like getting cobwebs all over your face as you walk through a door. The pleura over the lung is called 'visceral' and that over the chest wall 'parietal', but it is all the same stuff, a continuous layer. Parietal pleura is termed 'mediastinal', 'diaphragmatic', 'pericardial', depending on the area it covers. It lines the entire pleural cavity. Lungs fill the cavity to the extent that the cavity is reduced to a potential space. The visceral and parietal pleurae slide on one another in friction-free movement with each breath. The lubricant is a thin film of fluid diffused through the pleura.

The position of the fissures that divide the lungs into lobes, as represented on the surface of the chest, is important surface anatomy. The oblique fissure separating upper and lower lobes starts at the rib 6 costochondral junction, and extends obliquely up to the angle of rib 4. The horizontal fissure on the right, demarcating the superior border of the right middle lobe, extends along rib 4 from its anterior extremity to the point where the oblique fissure crosses rib 4. The triangular middle lobe lies between the two fissures. This surface anatomy is especially important in the accurate location of lung lesions as heard or seen on x-ray. Note that the lower lobe base is about two interspaces higher than the level of the pleural cavity base on expiration. The lungs normally do not completely fill the costophrenic recess. Thus, ribs 6-6-8-10 are the lung landmarks, not ribs 6-8-10-12.

The cartilage-ringed tracheal airway bifurcates into right and left main bronchi at the level of the angle of Louis. The point of bifurcation is known as the carina (L. keel). The main bronchi, also cartilage-ringed, enter the hila of the lungs in the medial midportions of their mediastinal aspects. Accompanying the bronchi are branches of the pulmonary artery. The lungs are divided up into anatomical segments, each of which is served by a bronchus and a pulmonary artery branch. Pulmonary veins leave the lungs at the hila but, within the lung, run between segments rather than in the segments, as do the arterial and bronchial branches. It is important that you understand the anatomical concept of lung segments. Segment removal based on this concept is possible. Each segment has a name but, at this stage, the precise names and locations of each segment (there are 10 on each side) need not be memorized.

The bronchi are nourished by small bronchial arteries arising from the descending aorta or by intercostal arteries.

Respiration. Rib-cage motions which alternately expand and contract the volume of the chest cavity have already been described. The diaphragm muscle when relaxed is dome-shaped but, when it contracts, the diaphragm shortens and flattens out. Thus, diaphragmatic contraction and relaxation will alternately enlarge and contract the vertical diameter of the pleural cavity. Putting this effect together with the effects of rib movements, it is clear that chest capacity is readily increased and decreased. With an increase in capacity, air rushes into the lungs through the open bronchial tree – inspiration. With relaxation and contraction in capacity, air is expelled – expiration. With inspiration, the lungs enlarge and slide down into the cleft between arching diaphragm and chest wall, the costophrenic recess. The lungs never completely open this recess, not even during deep inspiration. Thus, here the costal and diaphragmatic pleurae remain in contact. Remember that the base of the recess is the point of attachment of the diaphragm to the chest wall. The recess is deep and extends around the whole of each chest cavity and, whereas normally it is a potential space, it can become the hidden receptacle of considerable amounts (liters) of blood or other fluid which may be difficult to identify either on examination or on x-ray. It is also an important anatomical concept as with recognition of its depth comes the realization of how far down the pleural cavity extends.

Air has no business lying free in the pleural cavity. A hole in the chest wall or lung may release air into the cavity (pneumothorax) to cause marked interference with the normal physiology of respiration. Respiration depends on the production of negative pressure in the pleural cavity generated by rib and diaphragmatic activity. The lungs are elastic and usually considerably expanded by negative intrathoracic pressure. The thin layer of pleura-secreted serous fluid also sticks the parietal and visceral layers together while permitting easy sliding of one on the other (*cf* two beer-soaked playing cards one atop the other). When negative pressure is lost due to the entry of air into the pleural space, the lungs will collapse like the recoil of a released rubber band. Horizontal perforating wounds as low as ribs 9 and 10 may produce pneumothorax and respiratory difficulties (as well as intraperitoneal damage).

The diaphragm is innervated by the phrenic nerve, which descends from the neck across the cupola of the pleura, crosses in front of the lung hilum, and then down alongside the entire lateral vertical extent of the pericardium to reach and spread out over the diaphragm. The phrenic nerve is a mixed nerve, with motor fibers to the diaphragm, and sensory fibers from the pleura, pericardium and diaphragmatic peritoneum. Just as the heart descends from its embryological origin in the neck, the diaphragm, which also originates in the neck, descends into the future thorax, dragging the phrenic nerve down with it. The spinal origin of the phrenic nerve remains at C3–5. Injury to the phrenic nerve results in paralysis of the hemidiaphragm supplied. The hemidiaphragm becomes immobilized in a relaxed high arched position. The respiratory volume ('vital capacity') is significantly decreased. Irritation of the diaphragm and/or its pleural or peritoneal surface results in pain referred to the top of the shoulder. Can you figure out an embryological/anatomical basis for this? Note that the periphery of the diaphragm is formed from elements of the chest wall. Painful stimuli from the rim of the diaphragm are accurately felt at the site of inflammation. Embryological wisdom helps! (Skin at top of shoulder is innervated by C4).

Chest wall vessels. Segmental vessels nourish segmental areas. In the chest, these are termed 'intercostal' vessels. Accompanied by the intercostal nerves, they run in the costal groove in the undersurface of each rib. The arteries stem from two sources, the thoracic aorta and the internal thoracic arteries (branches of the subclavian arteries). The latter course subpleurally down the anterior chest wall about 1.5 cm away from the lateral border of the sternum, giving off intercostal arteries on the way. At the diaphragm, the internal thoracic arteries divide into musculophrenic and superior epigastric arteries. The musculophrenic follows the costal margin and continues to give off intercostal arteries; the superior epigastric continues down the anterior abdominal wall in the rectus muscle sheath and does the same job for the abdominal wall segments. A pericardiacophrenic branch of the internal thoracic artery is given off shortly after its origin from the subclavian and runs along the lateral pericardium with the phrenic nerve. The aorta gives off aortic intercostal arteries at the posterior extremity of each intercostal space. The clinical importance of these vessels is the potential for their injury and hemorrhage in any operation or injury involving the breast or rib cage. Furthermore, needle aspiration of the pleural cavity is frequently performed (thoracentesis) and you would be well advised to keep the needle away from the costal groove! Bleeding from torn intercostal vessels into the pleural cavity may be unsuspected until hemorrhagic shock becomes evident.

Venous drainage of the intercostal spaces is to the internal thoracic veins accompanying the internal thoracic arteries and to the azygos vein in the posterior mediastinum – a structure that will be described when that segment of the mediastinum is considered.

Mediastinum.   That part of the thorax that lies between the two pleural cavities, it comprises superior and inferior parts. The inferior area (plane of angle of Louis to diaphragm) is further subdivided into anterior, middle and posterior segments. Some structures traverse the entire mediastinum: the vagus and phrenic nerves, thoracic duct and esophagus.

Superior mediastinum.   Thoracic inlet (manubrium and T1, and connecting first ribs) to the level of the angle of Louis. Contains the aortic arch and arch branches, brachiocephalic veins, superior vena cava, trachea, thymus and left recurrent laryngeal nerve. It also contains the mediastinum-long esophagus, and vagi and phrenic nerves.

Inferior mediastinum

Anterior mediastinum.   Lies between the anterior surface of pericardium and the sternum; contains no significant structures. The two pleural cavities nearly meet here.

Middle mediastinum.   Filled with the pericardium and heart.

Posterior mediastinum.   Descending aorta and azygos system, esophagus, vagi and thoracic duct.

The posterior limit of the mediastinum is the anterior surface of the vertebral bodies. Thus, the sympathetic trunks running across the rib necks are not mediastinal structures.

Middle mediastinum: Pericardium.   This sac surrounding the heart and the cardiac ends of the great vessels rests on and is attached to the center and left paramedial parts of the diaphragmatic dome. It is applied above to the adventitia of the great vessels entering and leaving the heart, but the heart itself lies loose within the sac. A tough fibrous indistensible outer layer is covered by the pericardial pleura. The inner lining of the sac is a glistening slippery layer, similar to the pleura, which is reflected from the parietal walls of the sac onto the surface of the heart, starting at the site of pericardial attachment to the great vessels. As in the pleural cavity, there are parietal and visceral layers of pericardium and, between the two layers, a potential space in which the heart can move freely with each beat. Here also a film of serous fluid diffused through the serous pericardial layer is the lubricant.

Because of its indistensibility, the sudden collection of even a small amount of blood within the pericardial cavity (hemopericardium) will severely compromise the vital filling of the cardiac chambers. The resulting cardiac tamponade is life-threatening and must be relieved. A needle approach to the pericardium (pericardiocentesis) is anatomically safest via the costosternal junctional area, with the needle advanced through the base of the pericardial sac. When carried out properly, neither the pleura nor peritoneum are entered and there is no large coronary vessel in the vicinity. A direct approach through the parasternal anterior chest wall endangers pleura, lung, coronary arteries and internal thoracic vessels. (Chronic slow accumulation of pericardial fluid may slowly distend the pericardial sac to a massive size.)

Heart. First, check out the surface anatomy on yourself: Find the apical beat of the heart in the left fifth or sixth interspace, usually just medial to the midclavicular line. This strongest and most lateral beat is called the PMI – point of maximum impulse. The heart base extends from the PMI in a straight line medially to the xiphisternal junction along the dome of the diaphragm. The anterior base is the right ventricle. The PMI is produced by the apex of the left ventricle. The right end of the base is the lower end of the right atrium.

The top of the pericardium is at the level of the angle of Louis. The heart proper lies a good interspace below this. Bear in mind that the parietal pericardium is reflected some distance (2–3 cm) up the great vessels.

The right border of the heart is about 1 cm lateral to the right border of the sternum and is the right atrium.

The left border extends from the PMI up to the second intercostal space about 1 cm to the left of the sternal margin and consists of the left ventricle.

The bulk of the anterior surface of the heart is right ventricle and right atrium with a bit of left ventricle at the left border.

The posterior surface of the heart consists largely of left atrium and left ventricle.

The diaphragmatic surface of the heart is mostly left ventricle although, anteriorly, the right ventricle is present.

The heart does not stand squarely front to back, but is tipped upwards and rotated towards the left. The right heart is mostly in front with the left heart behind.

Structure. The right heart is subjected to less pressure than the left so the musculature is thinner. Both sides must pump the same amount of blood with each beat.

Right atrium. Filled by the inferior and superior venae cavae, and the coronary sinus returning blood from the myocardium proper, the right atrium contains the sinoatrial (SA) node. This invisible area, just below the entry of the superior vena cava, is the normal point of origin of the cardiac conduction system, the system that produces the coordinated progressive heart contraction. In the septum between right and left atria is the circular fossa ovalis, which marks the area of fetal bypass of the pulmonary circulation, the fetal foramen ovale. The bypass may remain unclosed at birth – patent foramen ovale – thereby producing unusual heart murmurs and, if large enough, circulatory embarrassment. The second component of the conduction system, also invisible on gross dissection, is the atrioventricular (AV) node. This area lies immediately superior to the membranous superior portion of the interventricular septum. The AV node relays the stimulus from the SA node to the ventricles. From the AV node, an AV bundle of specialized conductive muscle fibers (bundle of His) arises and splits into two bundles to run down both sides of the interventricular septum. In the right ventricle near its apex is a distinct trabecula, the septomarginal trabecula or marginal band. This band carries the right side branch of the bundle of His to the muscle of the lateral right ventricle; similarly, the left bundle conveys stimuli to the left ventricle. While the moderator band may be the only visible part of the system, knowledge of the general anatomical plan and function of the conduction system is important. The electrocardiogram (ECG) traces the passage of the electric conduction impulse through the conduction system. Cardiac problems which interfere with the normal transmission of the conduction impulse produce changes in the ECG. These changes may then be anatomically pinpointed. Although cardiac muscle is inherently capable of slow rhythmic contraction, such contraction is not controllable without the nervous system and a healthy conduction system controlled by the nervous system. The intact system enables sensitive, responsive, efficient and effective heart action.

Right ventricle.   Separated from the atrium by the tricuspid valve. Three valve leaflets are held in place by fibrous chordae tendineae attached to the valve margins and to three small cylindrical muscles (papillary muscles) arising from the ventricular musculature. The right ventricle exit valve is the pulmonary valve, which has three semilunar cusps beyond which the pulmonary artery extends.

Left atrium.   Situated posteriorly, the left atrium receives the four pulmonary veins – two from each lung. On the interatrial septum is the closed site of the foramen ovale. The left atrial exit valve is the bicuspid or mitral valve, held in place by chordae tendineae and papillary muscles. Posteriorly, the left atrium rests on the esophagus. If enlarged (heart failure, mitral disease), the esophagus becomes distorted.

Left ventricle.   Thick muscle wall. The mitral valve leaflets are held in place, as in the right ventricle, by chordae tendineae and papillary muscles. Note the membranous muscle-free portion of the interventricular septum just below the posterior and right aortic valve cusps. This area is a frequent site of interventricular septal defects, a congenital septal defect that may be life-threatening if not surgically corrected. The outlet valve of the left ventricle is the three-cusped aortic valve. In the sinuses of Valsalva, subtended by the left and right semilunar cusps, lie the openings of the left and right coronary arteries.

Knowledge of the surface anatomy of the heart valves is important as an aid to the interpretation of the sounds heard through the examining stethoscope (Figure 1.1).

Coronary circulation.   The coronary arteries arise as described above. The anatomy of the named major branches must be understood as they are so frequently diseased, and the site of disease can so accurately be delineated by use of current technology. Be familiar with the general course of the right coronary artery, nodal branch supplying the SA nodal area of the right atrium, marginal branch and posterior descending interventricular termination and, similarly, with the course of the left coronary artery with its anterior interventricular descending and circumflex branches. The latter meets the posterior interventricular branch of the right coronary artery in the posterior interventricular sulcus. The major coronary branches run around the heart in an atrioventricular sulcus (coronary sulcus), then down the heart in interventricular sulci both in front and in back.

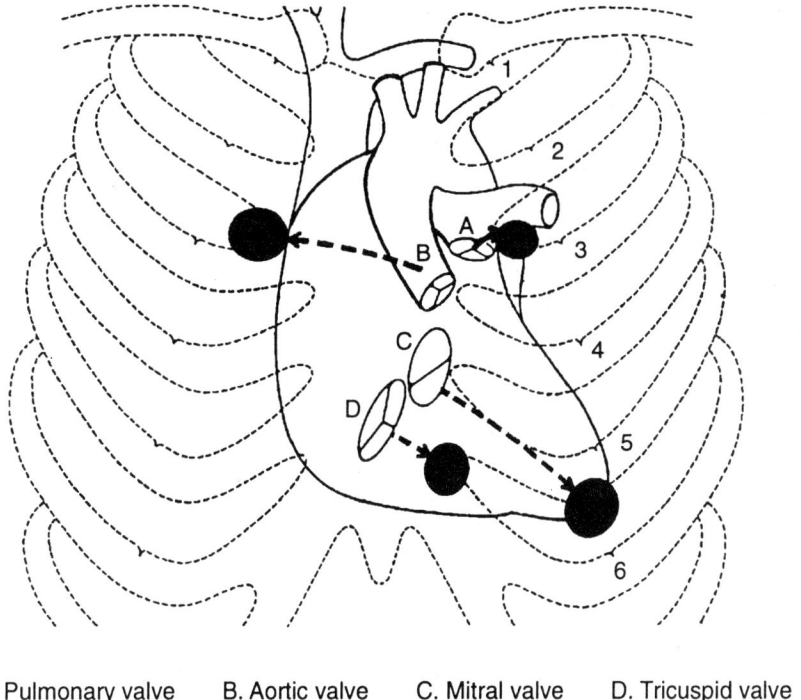

A. Pulmonary valve   B. Aortic valve   C. Mitral valve   D. Tricuspid valve

Figure 1.1 Surface anatomy of the heart valves. Arrows indicate direction of blood flow; black spots indicate the best points on the chest for auscultation of sounds produced by each valve

Coronary veins in general run with the arterial branches to enter the relatively large coronary sinus, situated posteriorly in the coronary sulcus between the left atrium and ventricle. The sinus empties into the right atrium between the opening of the inferior vena cava and the opening of the tricuspid valve.

Pulmonary vessels.

Pulmonary trunk and pulmonary arteries. Issuing from the right ventricle of the heart anterior to the origin of the aorta from the left ventricle, the pulmonary artery lies within the confines of the pericardial cavity clothed in visceral pericardium and bifurcates, after a 5-cm course, into right and left branches. The right branch passes posterior to the ascending aorta and superior vena cava, and in front of the right bronchus to enter the lung hilum. The left branch passes in front of the descending aorta and left main bronchus before entering the left lung hilum.

As the left branch crosses the aorta, another vestigial structure is found which represents the remnant of another fetal bypass of the pulmonary circuit. A stout fibrous band passes from the left pulmonary artery to the distal aortic arch – the **ligamentum arteriosum**. In fetal life, this was the ductus arteriosum (of Botallus). At birth, the ductus normally closes to allow pulmonary artery flow to reach the lungs. Occasionally, it remains patent and must be surgically closed.

Pulmonary veins.   Two trunks issue from each pulmonary hilus. These veins are the most inferior structures in the hilum. The right pair passes behind the superior vena cava; the left pair passes in front of the aorta. Both are caudal to the pulmonary arteries.

Superior mediastinum and posterior mediastinum.   These are considered together as many structures occupy both areas.

Aorta.

Aortic arch.   The relations of the aorta are important. It arises posterior to the pulmonary trunk within the pericardium covered with visceral pericardium. It arcs upwards anteriorly and to the right as the ascending aorta, and escapes from behind the pulmonary trunk to lie between the trunk and the superior vena cava, and in front of the right pulmonary artery. The aorta then arches from right to left well above the bifurcated pulmonary trunk, crosses the trachea, brushes along the side of the esophagus, and completes its arch alongside the vertebral bodies posterior to all of the left hilar structures. The aorta receives the ligamentum arteriosum at the conclusion of its arch as it passes behind the left pulmonary artery. At its apex, the arch reaches within an inch of the superior border of the manubrium at the level of the discs of T3–4. The beginning and ending of the arch are at the level of the angle of Louis. The branches of the ascending aorta are the coronary arteries. Branches of the arch are the entire blood supply of the head, neck and upper extremities; from right to left, these are the brachiocephalic, left common carotid and left subclavian arteries. These major vessels originate in the superior mediastinum and run upward and outward posterior to their accompanying veins, with no significant branches in the mediastinum.

Descending aorta.   Contained in the posterior mediastinum, its branches include the bronchial arteries, (posterior) intercostal arteries and multiple esophageal arteries. The descending aorta starts alongside the vertebral bodies at its origin at the end of the aortic arch and, by the time it reaches the diaphragm, is directly in front of the vertebral bodies. It lies behind the pericardium and esophagus, and passes through the diaphragm into the abdomen via the diaphragmatic aortic hiatus at T12 level (accompanied by the thoracic duct and azygos vein).

Great veins.   In the superior mediastinum, the two brachiocephalic veins run in front of the aortic arch branches and join immediately to the right of the ascending aorta to form the superior vena cava (SVC). The SVC courses alongside the ascending aorta to the pericardium, enters the pericardium and, shortly thereafter, the right atrium. Just before it enters the pericardium, the SVC receives the azygos vein.

The inferior vena cava passes through its hiatus in the diaphragm dome (T8 level) and almost immediately penetrates the basal pericardium. After a course of a few millimeters within the pericardial cavity, it enters the right atrium. The entire thoracic course of the IVC is approximately 2.5 cm.

Azygos system. This is the posterior system for collecting segmental blood flow. (The internal thoracic veins constitute the anterior segmental collecting system.) The azygos vein enters the thorax through the aortic hiatus, passes up the mediastinum to the T4 level, then arches forward cranial to the right lung hilum to enter the SVC just before the SVC enters the pericardium. Direct tributaries to the azygos vein are the segmental intercostal veins of the right thorax. A mirror-image of the azygos on the left receives the left thoracic intercostals. This is called the hemiazygos vein below midthoracic level and the accessory hemiazygos above this level. Both empty into the azygos through channels crossing the midline posterior to the aorta.

The azygos system also receives esophageal venous drainage. This becomes an important route for portal venous blood when the portal vein is obstructed. Chronic liver inflammation scarring (cirrhosis) heads the etiological possibilities. Obstruction is also relatively common with inflammatory disorders of the pancreas and with thrombus formation in the portal vein secondary to neonatal umbilical infection. Unfortunately, the azygos veins are too fragile to withstand the high portal venous pressure subsequent to these obstructing processes, and often enlarge in the esophagus or stomach and burst, producing torrential hemorrhage into the gastrointestinal tract. Such situational anatomy must be recognized and may be appropriately treated in many cases. Further discussion of this pathological situation is discussed in the Abdomen section.

Thoracic duct. This is the main lymphatic channel for all structures below the diaphragm, in the left chest, and in the left upper extremity, left side of the neck and head. The duct enters the thorax through the aortic hiatus and runs cranially along the anterior surface of the vertebral column between the aorta and the azygos vein. It veers towards the left in the superior mediastinum and enters the venous system in the neck at the junction of the left internal jugular and subclavian veins. There are many undissectable lymphatic channels which carry lymph from the right side of the supradiaphragmatic body to the junction of the right subclavian and internal jugular veins. The right and left systems communicate so that, if the thoracic duct is ligated or otherwise obstructed, these other lymphatic channels can easily handle the entire sluggish flow. If, however, the thoracic duct is pierced and not ligated, a profuse flow of lymph will demand control. A knowledge of thoracic duct anatomy is vital in such an event.

Nerves, sympathetic trunk and splanchnics in thorax.

Sympathetic trunk. The ganglionated bilateral sympathetic trunks have been described with the spinal nerves. Running subpleurally over the necks of the ribs, they are readily seen and are not, strictly speaking, mediastinal structures. The posterior border of the mediastinum is the anterior line of the vertebral bodies. The ganglionated sympathetic chain is responsible for the innervation of the vast amount of smooth muscle that invests every blood vessel, and of the innumerable sweat and sebaceous glands that inhabit every body segment. The chain supplies the postganglionic fibers that are contained in every spinal nerve and stimulate or inhibit these structures.

The viscera of the thorax and abdomen are not innervated by the spinal nerves. They are innervated by the autonomic system, but in a way that is anatomically different from that of the peripheral tissues.

19

Splanchnic nerves. The sympathetic supply to the abdominal viscera originates in the thorax as the splanchnic ('visceral') nerves. There are three nerves leaving the sympathetic trunk which run forward over the vertebral bodies anterior to the trunks to pass through the diaphragmatic, and greater, lesser and least splanchnic nerves. These preganglionics enter the sympathetic trunk as white rami from spinal nerves from T5–12, but pass through and out of the cord as 'splanchnic' nerves without synapse in the trunk ganglia. However, they synapse and become postganglionic nerves in special ganglia in the abdominal cavity. (Similar lumbar splanchnic nerves from white rami of the first two lumbar spinal nerves complete the sympathetic innervation of the abdominal, pelvic and perineal viscera.)

Vagus nerves. These important cranial nerves start in the brain and are responsible for the parasympathetic innervation of all the viscera in thorax and abdomen as far as the left side of the colon. In addition, the vagi have important laryngeal and pharyngeal functions. The right vagus appears in the thorax alongside the trachea, crossing superficial to the subclavian artery and deep to the brachiocephalic vein, to pass down the thorax posterior to the right hilar structures. The left vagus enters between the common carotid and subclavian arteries, also deep to the brachiocephalic vein, and passes over the left portion of the aortic arch and posterior to the left hilar structures. Both vagi then proceed caudally from the hilar level alongside the esophagus. The left vagus gradually works anteriorly so that, at the esophageal hiatus in the diaphragm, the left vagus becomes the 'anterior vagus', and the right becomes the 'posterior vagus'. The reason for this twisting will be obvious when the embryological rotation of the abdominal viscera is covered.

In the thorax, an important branch of the left vagus – the left recurrent laryngeal nerve – is given off at the arch of the aorta and runs back under the arch and ligamentum arteriosum to proceed cranially in the groove between the trachea and esophagus to reach the larynx. You should review the interesting embryological basis for this unusual anatomy, which involves the formation and development of the aortic arches. Injury to the recurrent nerve anywhere in its course results in paralysis of the left laryngeal muscles and a dramatic change in voice .

Nerve supply to thoracic viscera. These structures are supplied by the autonomic nervous system. As the heart developed in the cervical region, the airway budded off the gut tube in the cervical region and the bulk of the diaphragm also originated in the cervical region. Nerves running to these viscera, originally in the neck, were dragged down into the thorax with the viscera as they achieved their adult thoracic positions. This explains the long course of the phrenic nerves. In addition, the cardiac, pulmonary and esophageal nerves from both the parasympathetic vagi and the sympathetic trunks course down from the neck to reach these structures. When dissecting the neck, many of these tiny strands of nerve fibers can be seen running down to the thorax from the cervical vagi and cervical sympathetic trunks. Additional sympathetic nerves run to the thoracic viscera from the upper four thoracic spinal nerves and their sympathetic trunk ganglia. These are multiple and thread-like, and not readily dissected. The entire mass of supplying nerves from the neck and upper thorax forms a net-like cardiac plexus around the aortic arch, and spills over to form plexuses around the trachea and esophagus. This durable network can be demonstrated in the cadaver with ease. Branches from the plexuses control the cardiac conduction system, cardiac muscle, action of the smooth muscle and glands of the bronchial tree and esophagus, and smooth muscle of the visceral blood supply. They report sensory stimuli arising from these organs. Angina pectoris is the classic form of heart pain due to cardiac muscle ischemia. Like most visceral pain, it is poorly localized and diffuse, and is often referred to areas in the neck and extremities controlled by cervical spinal nerves of similar cervical cord segment levels.

Lymphatics of the thorax.   The lymphatics of the breast and gross anatomy of the thoracic duct have already been discussed. Gross evidence of the lymphatic system in the thorax other than the lymphatic duct can be appreciated by the presence of lymph nodes of varying size and number in the posterior mediastinum. They are usually most numerous and prominent in clusters around the lung hila, and bifurcation of the trachea and trachea proper. These are the so-called tracheobronchial nodes of the posterior mediastinal node group.

From the posterior mediastinal nodes, invisible lymphatics drain into the thoracic duct and into right-sided trunks that empty into the right subclavian–jugular junctional area.

Enlarged nodes (greater than a few millimeters in diameter) are the result of present or past inflammation or tumor. Once enlarged by an inflammatory process, a node may never again regain its previous smaller size. Note that enlarged nodes in the thorax may be secondary to inflammatory or malignant processes below the diaphragm that are draining upwards via the thoracic duct. Enlarged posterior mediastinal nodes are frequently visible in chest x-rays and are thus an invaluable aid to diagnosis. If large and 'vicious' enough, tracheobronchial nodes may knock out the left recurrent nerve, producing puzzling voice changes. The same nodes may enlarge sufficiently so as to visibly broaden the angle of the carina. The posterior mediastinal nodes constitute an important and sensitive anatomical area!

Esophagus.   A mere conduit in the thorax, its blood supply comes from aortic twigs and the draining azygos system. Its clinical importance is significant. It may harbor cancers that require removal without causing harm to important surrounding thoracic structures. It may be perforated by instruments or excessive vomiting and deposit hordes of mouth bacteria and foodstuffs into the mediastinal tissues or even the pleural cavities. The resulting inflammation requires emergency surgery. The esophageal hiatus in the diaphragm is at the level of T10. Hernias of the stomach through this opening are a frequent human frailty and promote the deposit of acid in the esophagus. Inflammatory reaction, narrowing and obstruction due to abnormal regurgitation of stomach acid is common. 'Heartburn' is a manifestation of stomach acid in the esophagus. (Normally it is quickly cleared.) Esophageal hiatus hernias often require surgical correction with meticulous restoration of normal anatomy. Finally, the bleeding from swollen esophageal veins seen in cases of portal vein obstruction has already been mentioned.

Thymus.   This final inhabitant of the superior mediastinum lies directly behind the manubrium. In the newborn, it is a large structure easily seen on chest x-ray. After puberty, the thymus involutes and is represented in the adult cadaver as a submanubrial pad of fatty tissue. In life, it has a slightly darker yellow color and is a little firmer than ordinary fat. In the cadaver, it is barely recognizable with any certainty.

As a final exercise, review x-rays and computed tomography (CT) scans of the chest. Pick out the shadows of the heart and great vessels, lungs, lobes and (sometimes) fissures. Identify structures such as the trachea, bronchi, pulmonary vessels and costophrenic recess. In addition, examine the embalmed lungs for impressions made by the aorta and its superior mediastinal branches, esophagus, diaphragm, vena cava and azygos vein. These exercises will help greatly in fixing the important anatomical locations and relationships in mind and will stand you in good stead in your clinical work.

Review Exercises

Sketch on the next four diagrams the surface projections of:

1. Attachment of the diaphragm
2. Location of lung fissures
3. Cupola of pleura

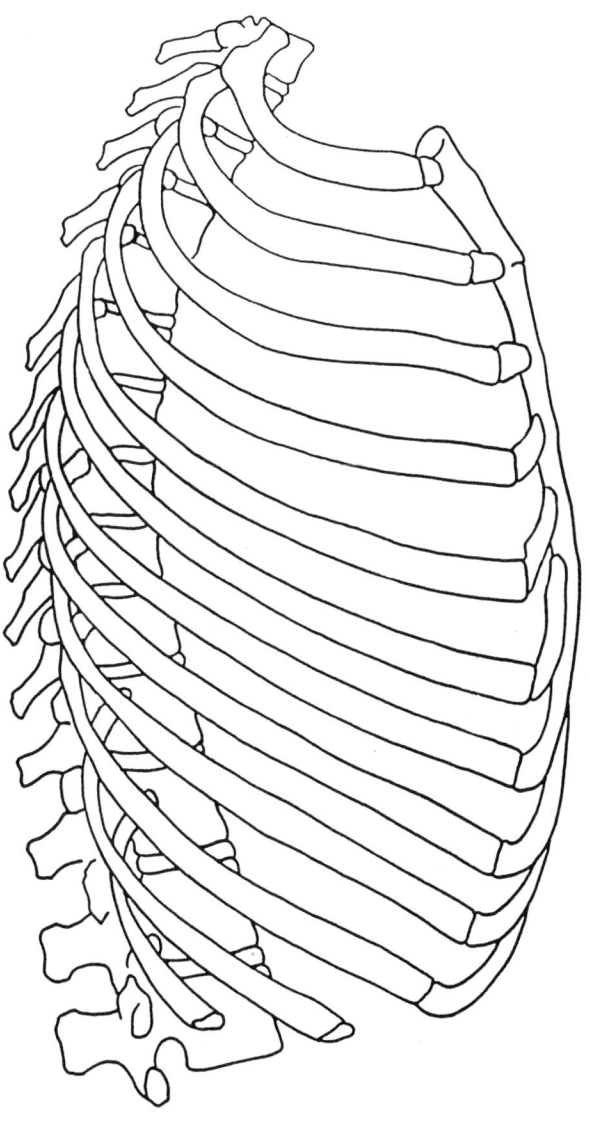

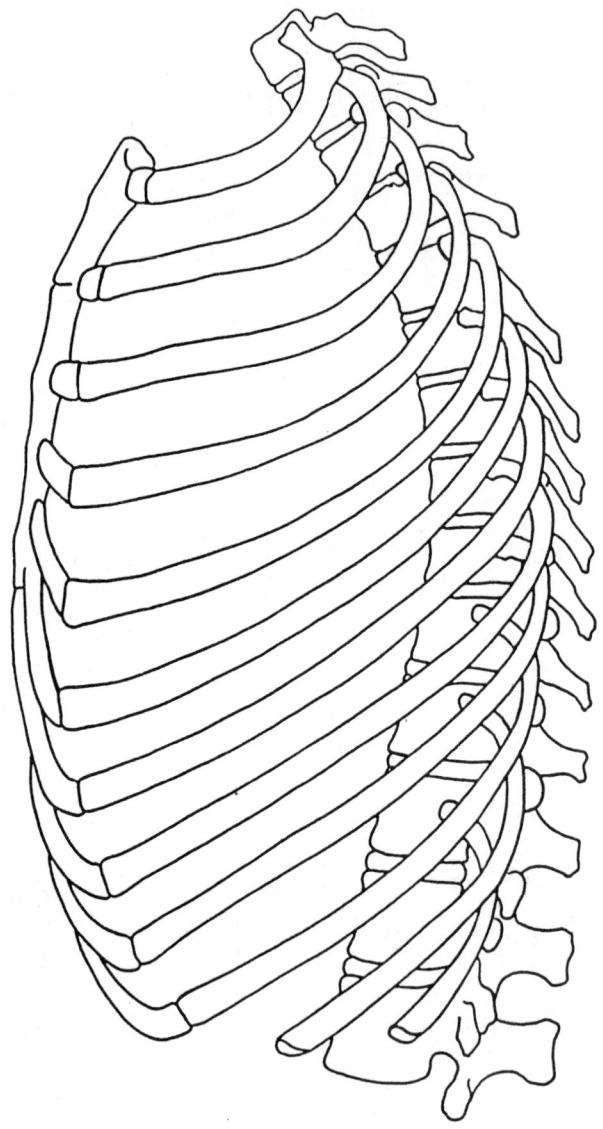

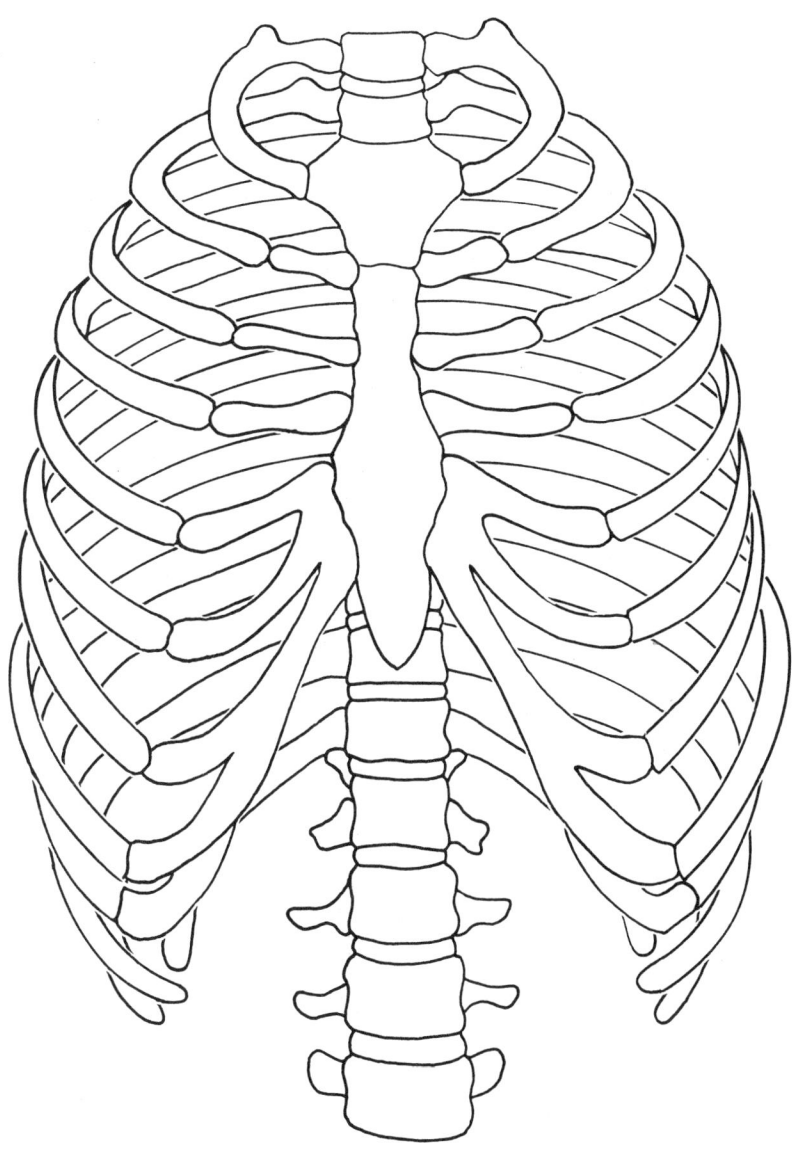

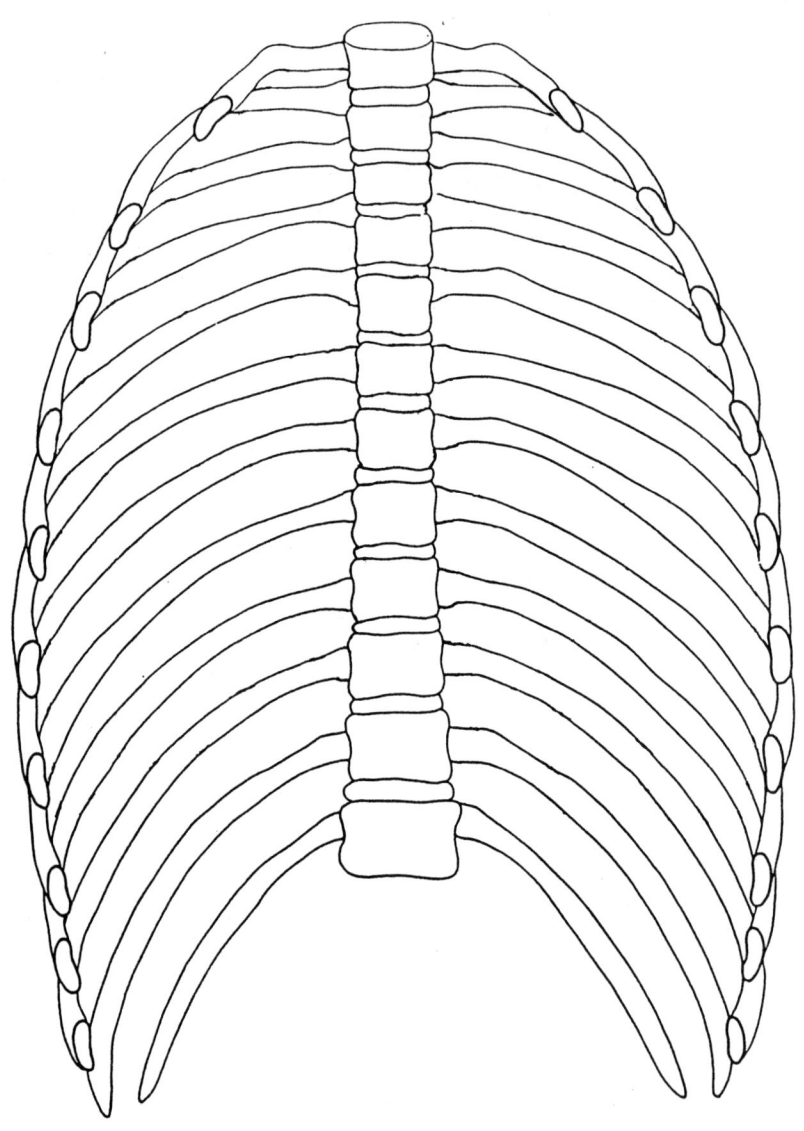

Sketch surface projections of:

1. Heart
2. IVC and SVC
3. Pulmonary trunk and aorta

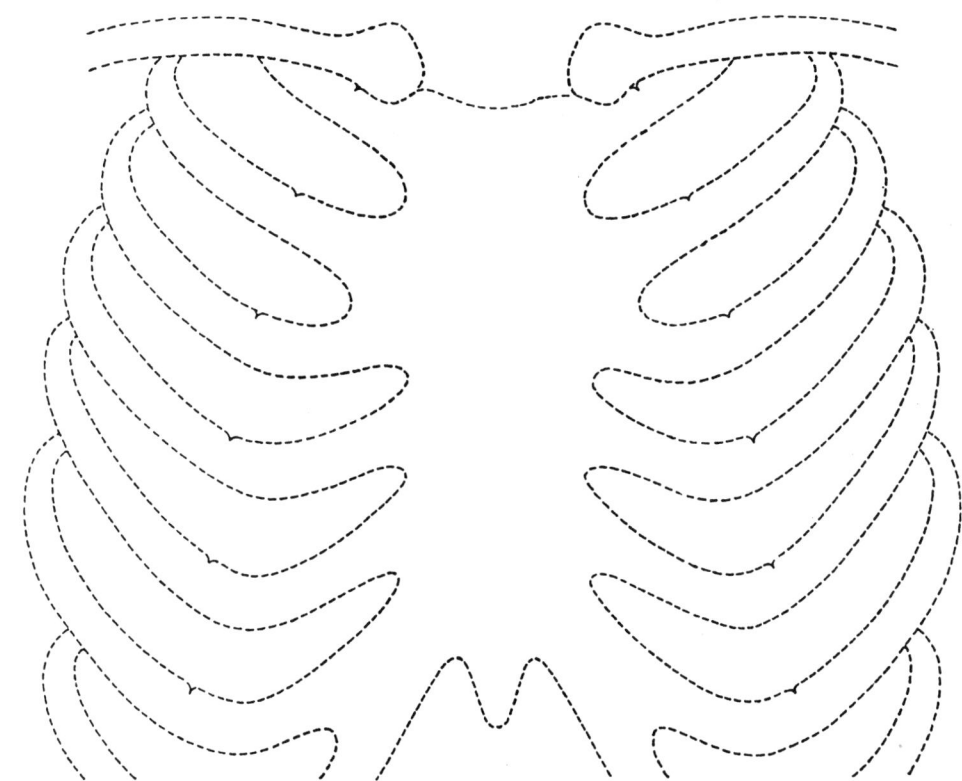

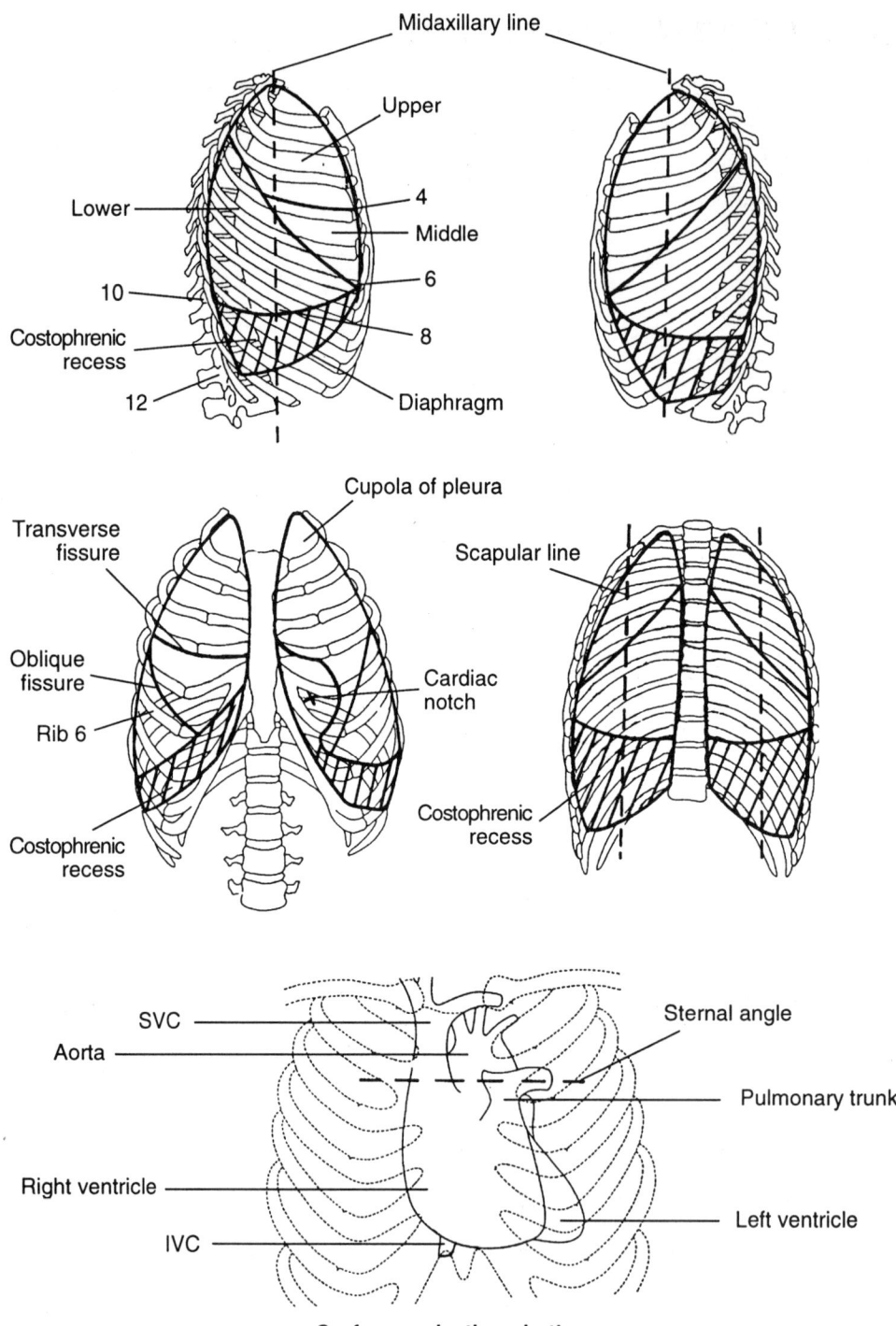

Surface projections in thorax

Note that the sternoclavicular joint is at the upper angle of the manubrium. The cupola of pleura (apex of lung) extends cranial to that level (clavicle)

Thorax Self-Assessment

Identify lettered items.

1. Articular portion of tubercle.

2. Rib neck.

3. Demifacets for vertebrae.

4. Angle of rib.

5. Costal groove.

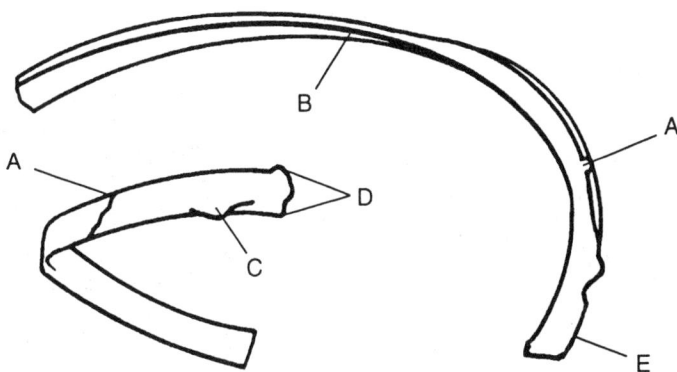

6. Identify the incorrect statement:

   A The breast is an organ within the confines of the skin.
   B The deep layer of subcutaneous fascia is separated from the pectoral muscle fascia by an areolar layer.
   C The lobes of breast glands are supported and defined by Cooper's ligaments extending from the deep layer of the subcutaneous fascia to the skin.
   D There are ± 20 lobes in the breast draining through ± 20 ducts at the nipple.
   E Lymphatic drainage from the breast is in its greatest part to the thoracic duct.

7. Regarding the spinal nerve, one statement is incorrect:

   A Each and every spinal nerve contains postganglionic sympathetic fibers.
   B There is one spinal nerve for each body segment.
   C Splanchnic nerves are composed of preganglionic sympathetic fibers.
   D Preganglionic sympathetic fibers for spinal nerve S2 travel to the S2 ganglion of the sympathetic trunk via the sympathetic trunk.
   E The sympathetic trunk extends from T1–S5.

8. Find the one incorrect statement:

   A   The left ventricle is filled through the mitral (bicuspid) valve.
   B   The coronary arteries are branches of the root of the aorta.
   C   Pulmonary venous flow passes into the right atrium.
   D   The left recurrent laryngeal nerve loops around the ligamentum arteriosum and the aortic arch.
   E   The bundle of His controls the rate of ventricular contraction.

9. Regarding the angle of Louis (sternal angle), choose the incorrect statement:

   A   Receives the third rib.
   B   Marks the level of junction of vertebrae T4–5.
   C   Overlies the bifurcation of the trachea.
   D   Lies caudal to the apex of the aortic arch.
   E   Defines the caudal border of the superior mediastinum.

10. In the presence of a total right chest pneumothorax, which one of the following statements is correct?

   A   Pneumothorax is always produced by a leak of air from the tracheobronchial tree.
   B   Free air in the right pleural cavity interferes with the ability to aerate the right lung.
   C   The mediastinum shifts to the side of the pneumothorax.
   D   The right diaphragm is found to be elevated.
   E   The lung is held in an expanded position due to the adherence of visceral and parietal pleura secondary to the usual viscid fluid film produced by the pleura.

11. Deep inspiration is aided by all of the following except one:

   A   External intercostal muscles.
   B   Pectoralis minor muscle.
   C   Serratus anterior muscle.
   D   Levator scapulae muscle.
   E   Anterior scalene muscle.

12. Regarding structures in the posterior mediastinum, choose the incorrect statement:

   A   The azygos vein enters the posterior mediastinum through the diaphragmatic aortic hiatus.
   B   The azygos system receives venous blood from the thoracic segments.
   C   Esophageal veins drain into the azygos system.
   D   The thoracic duct lies between the descending aorta and azygos vein.
   E   Ligation of the thoracic duct produces permanent lymphatic stasis and swelling of all structures below the diaphragm.

13. Regarding the heart and great vessels, choose the incorrect statement:

    A   The right atrium is directly related posteriorly to the esophagus. An enlarged right atrium will cause a significant deviation in the course of the esophagus.

    B   Portions of the right and left ventricles rest on the dome of the diaphragm.

    C   The descending aorta lies posterior to the esophagus.

    D   Pericardium covers approximately 3 cm of ascending aorta in the adult.

    E   The aortic arch ascends to the upper portion of the retromanubrial space.

14. A 6-inch knife blade penetrating the chest cavity (adult) through the left fifth interspace in the midaxillary line could injure all but one of the following:

    A   Left upper lobe of the lung.

    B   Left lower lobe.

    C   Inferior vena cava.

    D   Stomach.

    E   Phrenic nerve.

Answers and Explanations

1. **C** Anterior facet of the rib tubercle; for articulation with the transverse process of the lower of the two vertebrae with which the rib articulates.

2. **E**

3. **D** Articular facets of the head of the rib articulating with two vertebral bodies.

4. **A** For attachment of iliocostalis muscle.

5. **B** Houses intercostal vessels and nerve.

6. **E** Lymphatic drainage from the breast is primarily to the axillary nodes.

7. **E** The sympathetic trunk extends from the base of the skull to the coccyx. It supplies postganglionic sympathetic neurons in its ganglia, and postganglionic gray rami communicantes to all spinal nerves (C1–coccyx 1)

8. **C** Pulmonary venous flow is into the left atrium. It is called pulmonary **venous** flow because it is towards the heart. It is, however, full of oxygenated blood – quite arterial! The area of recurrence of the inferior laryngeal nerve is at two different levels on the two sides. The explanation lies in the disappearance of the fifth and sixth aortic arches on the right, and the persistence of a remnant (ductus arteriosus) of the sixth aortic arch on the left. On the right, the lowest persisting aortic arch is the fourth, the stem of the subclavian artery.

9. **A** The angle of Louis may be relied on to receive the second rib. The level of the angle is at the level of the 'crura' of the aortic arch, the ascending and descending portions. The apex of the arch is well up under the upper manubrium.

10. **B** Free air in the pleural cavity produces an intracavitary pressure greater than or equal to atmospheric pressure (greater than, if air can get in, but not out, due to a 'flap valve' of tissue at the leak site). The normal negative intrapleural pressure is lost, and the elastic lung will contract and collapse. As long as the leak remains open, respiratory efforts will result in outside air being sucked in which is then pushed out of the pleural cavity. There is much less resistance to air movement at the leak site than into the collapsed lung itself. Free air enters from holes in the chest wall as well as through leaks in the tracheobronchial tree or lung parenchyma. Mediastinal shift is to the side of more negative pressure, to the left in this case. The right diaphragm is flattened secondary to increased pressure on the right. The adhering fluid seal between parietal and visceral pleura is readily broken under the stress of atmospheric pressure within the chest cavity.

11. **D** All other choices involve muscles that can elevate the rib cage and thus increase lung capacity.

12. **E** There are sufficient collateral channel possibilities so that normal lymph flow is eventually restored.

13. **A** The left atrium lies on the esophagus and distorts the esophagus when enlarged. The pericardial sac encompasses the roots of the great vessels as well as the heart itself.

14. **C** The dome of the right diaphragm may extend as high as the fourth intercostal space. The fundus of the stomach is in contact with the underside of the dome. The interlobar oblique lung fissure runs from the costochondral junction of rib 6 up to the angle of rib 4. The knife may well penetrate either or both left lobes. It could reach the phrenic nerve running along the lateral aspect of the pericardium, but is not long enough to reach the IVC in an adult.

3 Abdomen

Abdominal wall. It is important and useful to relate abdominal wall features with vertebral and spinal nerve levels.

Xiphisternal = Spinal nerve T6 – Vertebral level T10
Umbilical   = Spinal nerve T10 – Vertebral level L4
Suprapubic = Spinal nerve L1 – Vertebral level is midsacrum

There is considerable sensory 'overlap' of sensory nerve fields.

The three flat abdominal muscles may be considered continuations of the three intercostal muscles. The ribs are gone from the human abdomen but, otherwise, the same plan applies. Thus, the external oblique proceeds from the ribs obliquely downwards and forwards (*cf* external intercostal). The internal oblique proceeds from the iliac crest and lumbodorsal fascia obliquely forwards and upwards to the ribs (*cf* internal intercostal). Transversus runs from the iliac crest, lumbodorsal fascia and lower ribs straight across to the midline (*cf* innermost intercostal). All three meet their contralateral counterparts in midline (linea alba).

The fourth abdominal wall muscle, the rectus abdominis, is the remains of a straight muscle running from pelvis to chin seen in lower vertebrates. (A second remnant is the series of strap muscles that persist in the anterior neck.)

The three flat muscles end in broad aponeuroses which form a fibrous sheath around the rectus abdominis before meeting in the linea alba. The structure of the rectus sheath has considerable clinical importance, especially to the surgeon. The external oblique aponeurosis lies entirely in front of the rectus. The internal oblique aponeurosis splits at the lateral margin of the rectus. The anterior split fuses with the external oblique aponeurosis in front of the rectus; the posterior split passes behind the rectus. The aponeurosis of transversus passes behind the rectus and fuses with the posterior split of the internal oblique to form the posterior rectus sheath. This arrangement, however, is interrupted below the umbilicus usually around the transverse level of the anterior superior iliac spines. Here, all aponeurotic layers pass in front of the rectus muscle; no aponeurotic posterior rectus sheath remains. The point of sheath change can usually be discerned as the 'arcuate' or 'semicircular line'. From here down to the pubis, the only fibrous tissue behind the rectus is the transversalis fascia. At the arcuate line, the inferior epigastric vessels are seen entering the rectus sheath. These vessels proceed upward to meet the superior epigastrics behind the rectus muscle within the upper part of the sheath. The lateral margin of the rectus sheath is visible as the 'semilunar line'.

The transversalis fascia is the fibrous lining of the abdomen lying between the muscle and peritoneum. It has significant strength especially in the lower abdominal-inguinal region. The anatomical role of this fascia in the inguinal region is described below.

Inguinal region.   The deep layer of the superficial fascia (Scarpa's fascia) is a significant and readily recognizable fibrous layer in the lower abdomen. It merges with the deep fascia – fascia lata – of the thigh caudally, which effectively closes off the subcutaneous space of the abdomen, an important anatomical feature as it confines extravasated urine from the penile urethra to the penis, scrotum and abdominal wall (described in more detail in the Perineum section).

The **pecten pubis** (pectineal line of pubis) is the ridge on the superior (horizontal) ramus of the pubis offering a line of origin for the pectineus muscle of the thigh. The **pectineal ligament** (Cooper's ligament) is attached to the pectineal line of pubis. The transversalis fascia is attached to this structure as it passes downwards to line the pelvis.

The **inguinal ligament**, extending from the anterior superior iliac spine to the pubic tubercle, is the thickened free caudal border of the external oblique muscle.

The **lacunar ligament** (Gimbernat's ligament) is the arcuate 'continuation' of the inguinal ligament running backwards – almost horizontally – and slightly downwards to merge with the pectineal ligament.

These three ligaments form a fibrous ring at the thigh crease for the femoral artery and vein, and lymphatics of the lower extremity .

(The bony parts and prominences of the pelvic bones are described in the Pelvis and Perineum section.)

Inguinal canal: Descent of testes.   The male gonad originates in the extraperitoneal tissue in the area of the developing kidney. To function properly, the gonad must reach the subcutaneous level. Undescended testes are unable to form sperm and may be more liable to undergo malignant change. Passage from extraperitoneal fat to the subcutaneous level requires penetration of the intervening layers of the abdominal wall: Transversalis fascia and abdominal wall muscles. The gonad is guided to its subcutaneous home by a fibrous cord, the **gubernaculum** (L. rudder), which is attached to the gonad and perineal skin (and eventually the scrotum). Throughout this passage, the gonad is accompanied by a fingerlike process of peritoneum, the **processus vaginalis**. The downward and medial path taken by the gonad and its traveling companion, the processus vaginalis, through the belly wall is the **inguinal canal**. On completion of the descent into the scrotum, the canal is filled by the **spermatic cord**, which contains structures that service the gonad and its duct. The canal thus extends from the level of the transversalis fascia to the level of the external oblique. In women, the gonad remains intra-abdominal. The female inguinal canal contains only the gubernaculum (renamed the ligamentum teres, which ends in the labium majus).

The first layer of abdominal wall penetrated by the gonad and its accompanying peritoneal sac is the transversalis fascia. The area of penetration is termed the **deep inguinal ring**. The transversalis fascia invests the testis, its accompanying peritoneal sac, and the train of service with one of its own layers, the **internal spermatic fascia**. The deep ring lies immediately cranial to the inguinal ligament and at around its midpoint level.

The next layer of abdominal wall musculature, transverus abdominis, is not penetrated by the descending testis. The muscle arises in its lowermost part from the inguinal ligament at a level cranial to the deep ring of transversalis fascia. These lowermost fibers arch over the spermatic cord to form a free caudal margin of muscle that eventually reaches an insertion in the pubic tubercle.

The gonad scrapes along under the internal oblique, in which the caudal-most fibers arise from the inguinal ligament in front of the deep ring. Fibers and fascia of this muscle adhere to the descending internal spermatic fascia-wrapped cord to form a second covering – the so-called **cremaster** (L. suspender) **muscle** and **cremaster fascia**, respectively. On completion of testicular descent in the mature male, loops of cremaster muscle fibers cover the cord down into the scrotum and then continue upwards to insert at the pubic tubercle as does the rest of the internal oblique. The cremaster muscle is innervated by the genital branch of the genitofemoral nerve. The testis is drawn upwards by contraction of the cremaster, a protective mechanism in cold weather and used by *Sumo* wrestlers to keep the gonads out of harm's way.

Guided medially and downwards by the gubernaculum, the testis passes through the aponeurosis of the external oblique and receives from this structure, *en passant,* a final covering, the **external spermatic fascia**. The region of perforation of the external oblique is immediately medial and superior to the pubic crest and is termed the **superficial inguinal ring**.

Having reached the scrotum, the testis invaginates into the peritoneal vaginal process in the same way that the lungs invade the pleural sac. Thus, the testis proper becomes tightly enclosed by a visceral layer of **tunica vaginalis** and rests in a sac of parietal tunica vaginalis. Two slippery surfaces are now in apposition so that the testis is free to move about (as are the lungs in the pleural cavity). The 'neck' of processus vaginalis extending upwards from the testis to the deep ring becomes absorbed and disappears, leaving the testis in the scrotum in its peritoneum-lined sac.

The inguinal canal thus stretches from the deep ring of transversalis fascia to the superficial ring in the external oblique aponeurosis.

Peculiarities of the transverse and internal oblique muscles must now be noted. The most caudal origins of both are, as stated above, the inguinal ligament. However, transversus arises lateral to the level of the deep ring whereas the internal oblique arises in front of the deep ring. Both muscles present a free caudal border. The two free borders join to form an arc of muscle and aponeurosis that extends, from their inguinal ligament origins, over the spermatic cord in the inguinal canal to insert on the pubic tubercle, pubic crest and medial portion of the pectineal line. The free-edged conjoined muscles and tendons (aponeuroses) form an arch called the **falx inguinalis** (L. sickle). The falx at its origin lies in front of the deep ring and the emerging cord, is cranial to the cord in the midportion of the inguinal canal, but lies behind the cord at the superficial ring level. Contraction of the conjoined muscle – tendon flattens the arch of the sickle to add a supporting abdominal wall layer over the inguinal canal and deep ring.

It is now apparent that the posterior (dorsal) back wall of the inguinal canal consists of the transversalis fascia attached to the pectineal ligament, and the medial end of the falx at its insertion. The two form the confining inner abdominal wall in this area. The outer covering of the canal is the aponeurosis of the external oblique. The canal is bound caudally by the inguinal ligament.

Spermatic cord and structures.

Maintenance structures are dragged down from above – from the original site of the testis, the region of the kidney – and extend to and from this level in the adult. They are:

Testicular artery, from the abdominal aorta;

Testicular veins, to the vena cava on the right and to the left renal vein on the left (embryologically, the 'left vena cava'). In the inguinal canal, these veins form a vine-like pampiniform (L. tendril) plexus;

Lymphatics, which ascend with the testicular vessels to para-aortic node chains;

Visceral efferent sympathetic nerves to the smooth muscle of the epididymis and vas, and the testicular glandular apparatus, which take origin at T10–11, and reach the testis and vas via splanchnics, renal plexuses and testicular vessels. Visceral afferent pain fibers retrace the efferent route to reach the spinal nerve, spinal cord and consciousness via rami communicantes. (Parasympathetic innervation may perhaps be present in the testes, although no anatomical evidence has yet been found.)

Vas deferens (L. vessel carrying away), the duct of the testis with its own little 'artery of the vas' derived from the superior vesical artery.

Supporting structures:

Internal spermatic fascia, from transversalis fascia;

Cremaster muscle and fascia from the internal oblique, together with its cremasteric artery from the inferior epigastric artery and its nerve from the genitofemoral;

External spermatic fascia, derived from the external oblique muscle.

Other anatomical features of the inguinal canal.

The back wall of the canal (transversalis fascia) is crossed on its underside by the inferior epigastric artery and vein which proceed diagonally upwards from the terminal external iliacs to the rectus sheath. These vessels divide the back wall into two areas: a medial **direct inguinal triangle** (Hesselbach) defined by the inferior epigastrics, inguinal ligament and lateral border of rectus sheath; and an area of canal back wall lateral to the inferior epigastrics containing the deep ring.

The iliohypogastric and ilioinguinal nerves (L1) from the lumbar plexus are segmental mixed nerves which, in their terminal courses, are sensory to the skin of the hypogastric suprapubic area and to the skin of the upper medial thigh, scrotum and root of the penis. Both may be seen in the inguinal canal as they emerge from between the transversus and internal oblique muscles. The ilioinguinal nerve is by far the more important clinically. It passes through the canal on the surface of the cord and exits with the cord through the superficial ring to reach its cutaneous destinations, and is at risk during inguinal area surgery. If injured, a very annoying numbness develops in the skin areas involved.

That is the whole intricate anatomical story of the inguinal region; and why bother with all this detail? Inguinal hernia, hydrocele, varicocele, undescended testis, among many others, are extraordinarily common clinical problems. These are problems that can be satisfactorily understood, and therefore satisfactorily treated, only if the embryology and anatomy of the parts are clearly understood. That's why!

Inguinal hernia. A hernia is "protrusion of a part or structure through the tissues normally containing it" (*Stedman's Medical Dictionary*). The specific hernias which concern us here are inguinal hernias. These involve the escape of intraabdominal structures – usually intestine – contained within a peritoneal sac into the inguinal canal. The normal tissues of containment for the abdominal contents are the peritoneum and the musculofascial abdominal wall. There are two 'normal' weak spots in the anterior abdominal wall. One is the gap through the transversalis fascia – the deep inguinal ring – made by the passage of the testis and the continued presence of the spermatic cord. The second is the virtually muscle-free inguinal triangle of Hesselbach. Hernias at these two areas are anatomically and clinically different. Deep ring hernias, called **indirect inguinal hernias**, are 'set up' by the normal process of the descent of the testis accompanied by the processus vaginalis of peritoneum. Hernias through the inguinal triangle are called **direct inguinal hernias**. These are 'breakdown products' of the tissues at this inherently weak area of the inguinal canal.

Indirect inguinal hernia. The processus vaginalis is a peritoneal diverticulum that is normally absorbed and therefore disappears during embryological development once the testis has been safely convoyed to the scrotum. It persists around the testis as the tunica vaginalis. The rest of it, between testis and deep ring, normally disappears. Total failure of this normal absorption results in a peritoneal pathway from peritoneal cavity to scrotum, a pathway that the slippery intestine will certainly find and enter. This produces a **congenital scrotal hernia.** If the processus is partially absorbed, but persists as a small peritoneal sac at the deep ring, a **congenital inguinal hernia** results. These congenital hernias are frequently seen. Most commonly, however, a tiny processus remnant or a small peritoneal dimple marks at birth the proximal vestigial end or point of obliteration of the fetal peritoneal process. Over the years, a 'new' processus may develop in response to the normal intra-abdominal stresses and strains of everyday life, and recapitulate, for a variable distance, the course of the original fetal processus. Intestine will then again find its way into this new sac and an 'adult' inguinal hernia appears.

In these hernias, the position of the peritoneal sac is always the same as was the fetal processus – **within** the fascial coverings of the spermatic cord. To find the sac, the surgeon must incise and reflect the cremasteric fascia and the deeper internal spermatic fascia of the cord. To cure the hernia, the sac must be opened, the contained intestine pushed back into the peritoneal cavity proper, and the sac itself ligated and excised flush with the plane of the peritoneal lining of the abdominal cavity. If the deep ring has been dilated by the bulk of the hernia, a few stitches in the transversalis fascia surrounding the deep ring will suffice in maintaining the snugness of the deep ring fascial borders around the spermatic cord, now restored to a normal diameter; otherwise, yet another 'processus' may form and a recurrent inguinal hernia ensue. These adult indirect inguinal hernias are classically diagnosed by invaginating the scrotal skin, and passing the examining finger up into the orifice of the external ring and distal inguinal canal. The hernia can be felt as a soft bulging mass that may push against the finger when abdominal pressure is increased by a cough or other straining maneuver.

Indirect inguinal hernias are dangerous and should be surgically repaired in almost all instances. External pressure from trusses and belts neither satisfactorily nor safely controls the exit of slippery intestine into the peritoneal sac in the inguinal canal. The neck of the peritoneal hernial sac is of a narrow diameter (at the deep ring). Intestine in the sac is often trapped there, like a finger in a bottle neck, and cannot be pushed back – 'reduced' – into the peritoneal cavity. In this case, the hernia is considered 'incarcerated' or 'irreducible'. It then requires only a little more pressure from the contents of the herniated intestine or twisting of the herniated intestine to obstruct the flow of intestinal content. This results in intestinal obstruction which must be relieved ('obstructed hernia'). Persisting obstruction results in increasing pressure in the incarcerated intestinal loop from accumulating mucosal secretion and vascular edema until interference with the trapped circulation is sufficient to produce gangrene ('strangulated hernia'). In the congenital types, the close quarters of the deep ring may produce enough pressure on the tiny prepubescent testicular artery to cause gangrene of the testis.

Direct inguinal hernia. Sharp rises in intra-abdominal pressure (cough, strain, etc.) continuously bombard the containing structures of the abdominal wall. The direct inguinal triangle is a weak spot where increases in intra-abdominal pressure become focused. With advancing age, obesity, lack of exercise and/or chronic straining (due to asthma, enlargement of the prostate interfering with the free flow of urine, chronic constipation, etc.), the abdominal wall tissues become fat-ridden, weak and stretched, and a localized bulging 'blowout' through a weak spot should not be surprising and frequently occurs. In contrast to indirect inguinal hernias, these direct hernias are due to acquired tissue weakness rather than developmentally based pathways through normal tissue. They are broad-based rather than narrow-necked, and the bulging hernia is directly through the abdominal wall and anatomically has nothing to do with the spermatic cord *per se*. Furthermore, unlike indirect hernias, the enlarging direct hernia is not led down into the scrotum, a frequent event in neglected acquired indirect inguinal hernias which recapitulate the embryological course of the processus vaginalis. Repair of direct inguinal hernias is indicated to control discomfort and the inevitable gradual increase in size. Repair may require artificial support for the abdominal wall in the form of grafts of various materials (and attempts to correct the factors responsible for the hernia).

Hydrocele. Fluid distention of the peritesticular tunica vaginalis cavity (often associated with a persistent tiny vestigial connection with the parent general peritoneal cavity) presents as a painless translucent swelling in the scrotum. The cause of the excess fluid is not known. Repair consists of resection of the parietal layer of tunica and removal of any persisting remnant of the processus vaginalis.

Varicocele. (L. *varix* = dilated vein.) Swelling and tortuosity of the veins of the pampiniform plexus may produce an uncomfortable and unsightly swelling of the spermatic cord. This occurs usually on the left side perhaps due to some obstruction of flow at the right-angle junction of the left testicular and left renal veins, or due to obstructing pressure of the superior mesenteric artery as it crosses the left renal vein. Varicocele is surgically correctible by resection of excess venous tissue in the cord.

Undescended testes. Having failed to reach the scrotum, one or both testes may reside in the inguinal canal or remain intra-abdominal. Often, under the influence of natural or administered hormones, the testis will achieve the normal scrotal position as puberty approaches. Otherwise, an attempt must be made to bring the testis down surgically. The undescended testis is not able to produce sperm and is thought to be vulnerable to the development of testicular cancer.

Inguinal hernia in the female.   The female inguinal canal has the same general origin and structure as the male, but with the obvious difference that the female gonad remains intra-abdominal rather than traversing the canal. A processus vaginalis and a gubernaculum are formed, the latter found as the **round ligament** running through the canal to the labium majus – the homologue of the male scrotum. (Proximally, the round ligament passes through the broad ligament to the uterus and then to the ovary.) Because the canal is 'unused', contracted, and contains no space-occupying spermatic cord, the direct triangle is small and well protected by overlying muscle. Direct inguinal hernias are almost never seen in women. Indirect inguinal hernias do occur, though not with the frequency seen in men. A perisisting processus (termed the 'canal of Nuck' after a 17th-century Dutch anatomist) may house a loop of intestine. Occasionally and not surprisingly, in congenital female hernia, the ovary is found in the inguinal canal. These hernias are readily repaired by the restoration of contained structures to the peritoneal cavity where they belong and by resection of the persisting processus vaginalis.

Abdominal cavity.   This is totally lined by parietal peritoneum, thus forming a **peritoneal cavity**. Developing in, and suspended from anteriorly and posteriorly attached mesenteries, are the organs of the GI tract and the spleen. Mesenteries are double-layered folds of peritoneum extending from the parietal peritoneum. Their suspended viscera are covered by the mesenteric peritoneum. The same stuff is called parietal (lining the abdominal cavity) or visceral (covering the organs) peritoneum, depending on its location (*cf* pleura, pericardium).

The embryological dorsal mesentery extends the full length of the peritoneal cavity, attached for its entire length to the posterior midline. The embryological ventral mesentery extends from the anterior parietal peritoneum to the foregut only. Thus, in the developed human, it is seen only as far down as the midportion of the second part of the C-shaped duodenum. Here its termination is seen as the free lateral border of the hepatoduodenal ligament. The foregut structures then are suspended between the dorsal and ventral mesenteries. The mid- and hindgut have only the dorsal mesentery to keep them warm. Remnants of the ventral mesentery include the peritoneal falciform ligament and the remains of its once important umbilical vein, now found as a tubular round ligament in the falciform free border.

Tracing the two peritoneal layers of the falciform cranially and then medially, the two layers open up to embrace the liver, except for the bare area of the liver. They meet on the other side of the liver to form the two-layered gastrohepatic or lesser omentum and the free-bordered hepatoduodenal ligament. The latter contains the hepatic artery, portal vein and bile duct between its two layers. Remember, the liver, gallbladder and biliary tree develop in the ventral mesentery from a foregut duodenal bud. The two layers of gastrohepatic omental peritoneum open up to clothe the front and back of the stomach from the lesser to the greater curvature, then meet again to form what is now the dorsal mesentery. Between stomach and spleen, the dorsal mesentery is called the gastrosplenic omentum. After encasing the spleen, the two layers meet again and pursue a backward course as the splenorenal ligament to reach the posterior parietal peritoneum.

The greater omentum is a peculiar overgrowth of the dorsal mesentery from the greater curve of the stomach. It extends downwards as a double-layered fatty peritoneal apron to and over the transverse colon (gastrocolic omentum), continues downwards for a variable distance, then doubles back again over the transverse colon. The two-layered greater omentum, as it descends and then ascends, forms a four-layered mesentery with a space between the descending and ascending layers. The four layers eventually fuse to eliminate the space, resulting in a two-layered peritoneal apron containing some fat, blood and lymphatic vessels. It then fuses with transverse colon dorsal mesentery to reach the posterior parietal peritoneum fused to this structure. The dorsal mesentery of the midgut is anatomically more straightforward and is described in the Midgut section.

Mesenteries contain and convey blood and lymphatic vessels, and motor and sensory nerves, servicing their subtended viscera. The celiac axis to the foregut, the superior mesenteric to the midgut and the inferior mesenteric to the hindgut are the servicing arteries. Similarly named portal venous tributaries run together with these arteries. Nerves ride on the backs of the arteries to reach the viscera. Lymphatics also travel along with the main vessels. Without mesenteries, there would be no means by which these servicing elements could reach their targets. The ventral mesentery, which *in utero* contained the umbilical vein, in the adult contains its remnant, the ligamentum teres of the liver in its falciform ligament. Not surprisingly, by maintaining its embryological origin, the falciform extends from the umbilicus to the liver attached to the anterior abdominal wall. Between the liver and stomach-duodenum, the ventral mesentery contains the servicing vessels, nerves and ducts for the structures that have developed within it – the liver, gallbladder and biliary tree – reaching the liver area via the accessing hepatodudenal ligament. Vessels and nerves to the lesser curve area of the stomach are also contained in the ventral mesentery (gastrohepatic omentum) and reach the stomach and gastrohepatic omentum via a fold of peritoneum at the cranial end of the lesser curvature. Dorsal and ventral mesenteries meet at the stomach.

Embryological review helps in understanding of details of distribution of the peritoneum, which has a definite clinical relevance. The foregut (stomach and proximal duodenum), together with the 'tributary' pancreas, biliary tree and liver plus spleen, all develop within the peritoneal leaves of the dorsal and ventral mesenteries. The lesser sac represents the distorted right upper space of the embryological peritoneal cavity, as follows (Figure 3.1). Growth of the liver into, and then with, the septum transversum (the diaphragm anlage) does away with peritoneum here, resulting in a peritoneum-free bare area of the liver (and diaphragm). The ventral mesentery continues on to the lesser curve border of the stomach but, in the latter's growth, a twist occurs wherein the greater curve flips laterally to the left and anteriorly so that the ventral mesentery-stomach body-dorsal mesentery 'sheet' lies in a side-to-side frontal plane rather than in a front-to-back sagittal plane. The dorsal mesentery is in a frontal rather than sagittal plane and runs from the greater curve of the stomach to the spleen. After encasing the spleen, it curves backwards to reach the posterior parietal peritoneum from where it started. Thus, a sac has been formed. The liver grows to occupy most of the cranial part of the ventral mesentery, leaving only a small right lateral-facing opening to the sac, the foramen of Winslow.

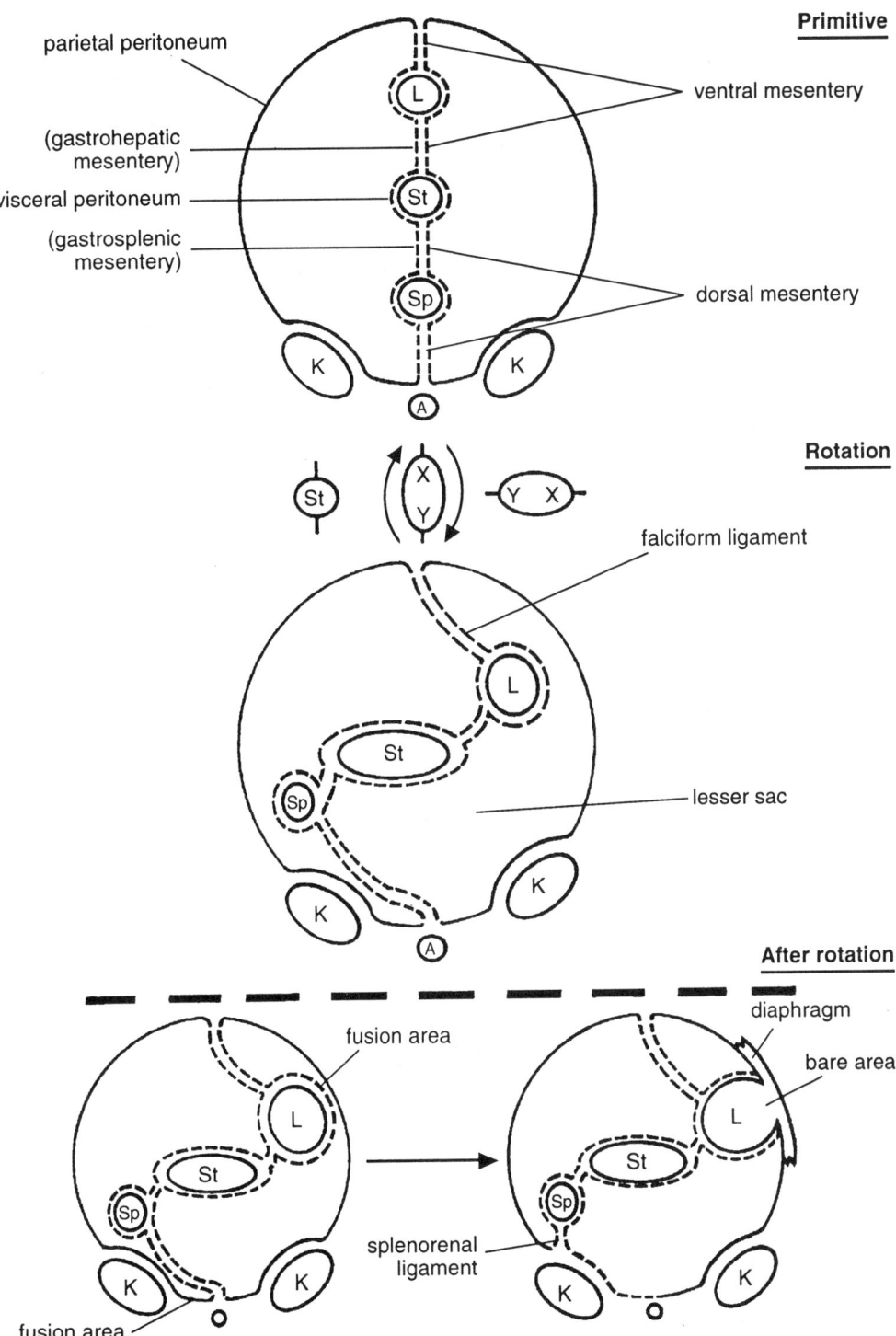

Figure 3.1 Foregut rotations, and formation of the lesser sac, splenorenal ligament and bare area of liver

Remember the propensity for two juxtaposed layers of peritoneum to fuse and obliterate. Such apposition occurred when twisting of the stomach laid the dorsal mesentery, and its contained duodenum and pancreas, against the posterior parietal peritoneum. The posterior apposed mesenteric layer and parietal peritoneum fused and disappeared. The duodenum and pancreas are thus fixed posteriorly and covered by the remaining layer of dorsal mesenteric peritoneum. They are now considered 'extraperitoneal' although, as can be seen, this is an artificial designation. The fusion area presents an easily entered bloodless plane which, when surgically developed, restores pancreas and duodenum to their natural intraperitoneal position.

A large part of the dorsal mesentery of the spleen similarly fuses with parietal peritoneum over the upper left kidney. A short splenorenal peritoneal ligament is all that remains of the dorsal mesentery between spleen and posterior parietes. The fusion plane between dorsal mesentery and posterior parietal peritoneum at the posterior termination of the splenorenal ligament can also be surgically developed bloodlessly back to the midline where the dorsal mesentery artery originates (celiac axis). The spleen itself remains fully intraperitoneal, with visceral peritoneum of the dorsal mesentery still surrounding it. Posterior fusion fixation of the proximal duodenum narrows the opening to the lesser sac to a small orifice between it and the bare area of the liver – the epiploic foramen of Winslow.

The roof of the foramen is still ventral mesentery (hepatoduodenal ligament) and carries in it the vessels, nerves and ducts of the liver and its biliary tree. The posterior wall of the foramen is posterior parietal peritoneum (overlying the inferior vena cava). So, on entering the lesser sac through the foramen of Winslow, the pancreas can be seen lying posteriorly under peritoneum with the back wall of the lesser sac stretching from just within the foramen to the splenic hilum. Anteriorly lie the portal vein, hepatic artery and common bile duct in the hepatoduodenal ligament. Farther to the left, the gastrohepatic ligament to the lesser curve of the stomach, the stomach itself and the dorsal mesentery from the greater curvature to the spleen (gastrosplenic ligament) form the anterior cover of the lesser sac. The splenorenal peritoneal ligament closes the sac in its left lateral aspect.

The greater omentum is simply an excessive growth of the midportion of the dorsal mesentery from the greater curve of the stomach down over the transverse colon. It then doubles back to recross the transverse colon and finally reach the posterior abdominal wall. Again, peritoneal fusion occurs between the returning omentum and the transverse colon and its mesentery. Caudal to the transverse colon fusion, the two two-layered walls of the greater omentum fuse into a single double-layered structure. Thus, the lesser sac caudal to the stomach has, as its anterior wall, the gastrocolic portion of the greater omentum. The caudal wall is made up of the fused returning greater omentum and transverse mesocolon. The transverse position of the transverse mesocolon is obviously the result of the embryological rotation of the colon (see Figures 3.1 and 3.2).

Blood vessels to this convoluted foregut still originate in the posterior midline and run to their appointed destinations either in the 'retroperitoneal' areas (for example, splenic and gastroduodenal arteries) or in the still existing mesenteries (for example, short gastrics, right and left gastric, and hepatic proper). Fused areas may be entered and bloodless cleavage planes developed back to original midline positions. The posterior aspects of the pancreatic head and the duodenal C can readily be seen by incising the peritoneum sweeping off the duodenum and bluntly dissecting medially until the midline-

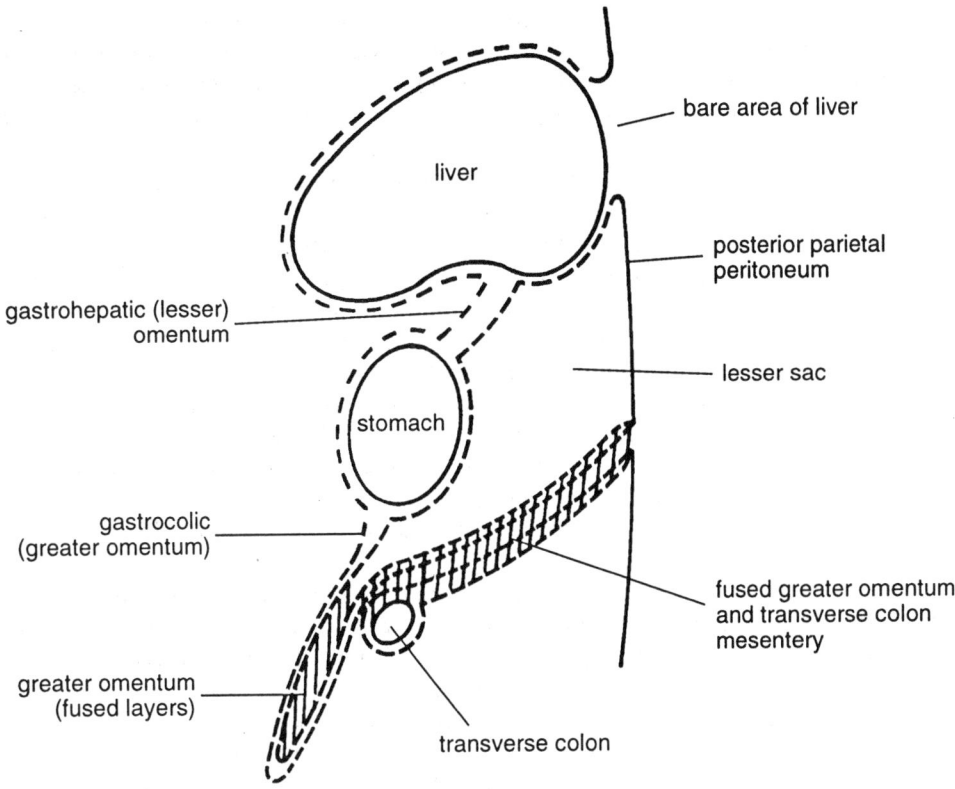

Figure 3.2 Sagittal section: Median plane (broken lines indicate visceral peritoneum)

originating celiac and superior mesenteric arteries intervene at their aortic take-offs and tether the two. Similarly, the spleen, and tail and body of the pancreas can be restored to their primeval midline position by incising the peritoneum of the splenorenal ligament and bluntly dissecting medially to the midline behind the pancreas. No crossing blood vessel is encountered until the artery of the foregut is reached – the celiac axis. This is where it originally entered the dorsal foregut mesentery.

Embryologically based anatomical facts are used daily by knowledgeable surgeons to render otherwise difficult operations on 'retroperitoneal' foregut structures safe and simple. The deep-lying spleen, for example, can be mobilized and brought, still attached, all the way out of the abdominal cavity using this embryologically based maneuver.

The embryology and resulting peritoneal disposition of the mid- and hindgut are also of interest and useful to know. Remember that the growth of the midgut results in a physiological hernia through the umbilicus of the fetus. In the physiological hernia sac, the midgut loop rotates about its superior mesenteric artery axis in a counterclockwise direction (with the clockface looking out from the abdomen). This drags the distal or postarterial segment of midgut around and across the proximal prearterial segment. The superior mesenteric vessels will now lie across (ventral to) the distal duodenum as the rotated gut returns to an intra-abdominal locus. This relationship is important in adult anatomy. First, it facilitates finding the superior mesenteric vessels even in a grossly obese belly. Second, a vomiting syndrome occurs and can be explained, when displayed on x-rays, by an unusual degree of pressure on the duodenum caused by the crossing vessels. Termed the 'superior mesenteric artery syndrome', surgical correction is usually required.

The midgut, which forms the proximal half of the large intestine, and the hindgut participate in this growth and rotation. The counterclockwise torque drags the ileocecal junction and colon upwards along the left side of the abdomen, then across and finally down into the right lower abdomen. The large intestine has now assumed an inverted 'U' shape. The mesenteries (visceral peritoneum) of the two arms of the U (ascending and descending colons) rest against the parietal peritoneum in their new positions. As with more proximal rotations, the now apposed parietal and visceral peritoneal layers fuse and disappear, leaving the ascending and descending U limbs fixed to retroperitoneal tissues and thus apparently 'retroperitoneal'. The embryologically aware dissector-surgeon can apply this knowledge when operating on the colon itself or on structures of the retroperitoneum hidden by the attached large bowel. By incising the peritoneum reflected off the colon onto the parietes, mobilization of both ascending and descending colons is readily accomplished bloodlessly, returning them to their original midline position, complete with their feeding vessels and nerves from the posterior midline. This is where they orginally entered the dorsal mesentery of the mid- and hindgut (superior and inferior mesentery vessels). The retroperitoneum lies completely exposed. Thereafter, the colon limbs may be dropped back into their normal positions where they reattach themselves in short order.

The mesenteries of the transverse and sigmoid colon segments remain free. The most distal portion of the large intestine disappears through the parietal peritoneal floor of the abdominal cavity to become the truly extraperitoneal rectum.

The umbilicus is a beehive of embryological activity. The umbilical vein running through it is represented by the ligamentum teres of the liver contained in the falciform ligament of the formed newborn. The umbilical arteries are seen in the adult as two peritoneum-covered fibrous ridges sweeping off the dome of the bladder after giving off their superior vesical branches and running up to the umbilicus. These adult embryological remnants are termed the **medial umbilical ligaments**. They are of no clinical significance except as landmarks. In the midline lower abdominal wall lies a single peritoneum-covered cord, the remnant of allantois called the urachus, termed the **median umbilical ligament**. The allantois is the primordial excretory duct of the fetal hindgut, passing into the body stalk and fetal chorion. With division of the hindgut (cloaca) by the urorectal septum, the allantois becomes involved in the formation of the bladder. The allantoic urachus normally completely disappears except for the median ligament. Occasionally, the lumen of the urachus remains patent and a discharge of urine then appears at the umbilicus. This patent urachus must be removed. (The inferior epigastric vessels form peritoneum-covered ridges – the **lateral umbilical ligaments** – as they pass upwards to enter the rectus sheath at the arcuate line.)

The connection between the gut tube and the yolk sac (from which the gut developed) is known as the omphalo(umbilical)mesenteric duct. This too may fail to close and a patent duct in the newborn will emit a fecal discharge through the infant navel. This connection must also be recognized and surgically removed. More commonly, a remnant of this duct may persist at its junction with the midgut (usually the terminal ileum). This diverticulum (Meckel's) may or may not cause trouble (bleeding, obstruction, perforation) and require removal. Other omphalomesenteric duct remnants that may be seen as the result of incomplete resorption include mucus-secreting pits in the umbilicus, a fibrous cord reaching from ileum to umbilicus and cystic structures in the course of such a cord.

Finally, remember that the physiological herniation of the gut is into the umbilicus. Persisting umbilical herniation of any degree may be seen in the newborn. Thus again, embryological knowledge helps to clarify many otherwise puzzling neonatal-infant-childhood findings.

In the human abdomen, the base of the small bowel (midgut) mesentery is found attached to the posterior abdominal wall in a diagonal line from the duodenojejunal junction to the ileocecal junction – a 15 to 20-cm root in the adult. The diagonal position is subsequent to the midgut rotation. This 15-cm root services15-20 feet of small intestine by means of the ruffled fan-shaped growth of this part of the dorsal mesentery.

Also note that there are two deep pockets in the peritoneal cavity. The first is the space around the upper right kidney and suprarenal gland dorsolateral to the duodenum and the inferior vena cava, behind (dorsal to) the liver and backed by the peritoneum covering the caudal diaphragm. The space is roofed by the lateral aspect of the hepatic coronary ligament, and the right lobe and right triangular ligament of the liver. This is called the hepatorenal recess (Morison's pouch). The second is the rectovesical pouch (male) or rectouterine pouch (of Douglas; female), formed where posterior parietal peritoneum sweeps downwards and forwards to cover the bladder or uterus as it progresses to the anterior abdominal wall. These two recesses are the most dorsal parts of the peritoneal cavity and clearly can harbor puddles of fluid in the recumbent patient. They are frequently loci of intraperitoneal abscesses. Also note that the fixed right colon forms the medial wall of a peritoneal gutter, a fine channel for fluid movement between the two recesses. The descending colonic gutter on the left also connects a left subphrenic space with the pelvic cul-de-sac of Douglas. Pus from Morison's pouch will inflame the diaphragm and overlying pleural cavity. If neglected, such an abscess may rupture into the lung, and drain via the bronchi and mouth.

Foregut organs of the abdomen.

Abdominal esophagus
Stomach
Duodenum to the region of the ampulla of Vater
Pancreas
Biliary tract
Gallbladder
Liver
(Spleen)

The arterial blood supply to all comes from the old foregut artery in the dorsal mesentery – the celiac axis. Venous drainage is to the liver via the portal system. Lymph drains to the cisterna chyli (chyle = lymph) in channels running alongside gut vessels and passes through innumerable filtering lymph nodes on the way. Neural management is handled by both segments of the autonomic system via the vagi and the thoracic splanchnics.

Abdominal esophagus. Approximately 2 cm of terminal esophagus resides in the abdomen from the diaphragmatic esophageal hiatus, level T10, to its junction with the cardia of the stomach. The anterior (left) and posterior (right) vagi lie along the front and back of the esophagus. These nerves may be single trunks or multiple strands. The esophageal hiatus is encircled by fibers of the diaphragmatic crus which tend to support and, perhaps to a degree, sphincter the esophagus. The terminal esophageal circular muscle, although not remarkable anatomically, has a much more important sphincter role and is called the 'lower esophageal sphincter'. (The 'upper esophageal sphincter' involves the cricopharyngeus portion of the inferior constrictor of the pharynx.) Vagus nerves can (must) be completely visualized in operations designed to eliminate neural stimulation of gastric acid production. Division of the nerves – vagotomy – is a frequently performed procedure. Incomplete division will not satisfactorily eliminate neural acid production and the operation will fail. All strands must be identified and dealt with. (Modern surgical techniques spare branches of the vagi not involved in acid production; see Stomach below.)

The venous drainage of the esophagus is to the low-pressure azygos system. The terminal esophagus lies between the drainage beds of the azygos and portal systems. The two communicate here. When portal pressure rises due to obstructing disease in the liver, or thrombotic occlusion of the portal or hepatic veins, portal pressure rises and portal blood will seek exit through the connecting low-pressure azygos system. However, the thin-walled azygos vessels cannot comfortably carry the flood of high-pressure portal blood, and will balloon out (esophageal varices) and frequently rupture, releasing torrential hemorrhage into the esophagus and stomach. Portal hypertension with esophageal varices and rupture is a common medical emergency situation. With good fortune, the incident can be controlled by portal pressure-reducing drugs, thrombosing injections of the esophageal varices or by operations that shunt portal blood to systemic channels by surgical bypass anastomosis of large portal branches to the vena cava or one of its large tributaries, or by intrahepatic shunt tube placement. Large systemic veins can easily withstand the high portal pressure.

Hernias of the stomach up through the esophageal hiatus are common. They stretch and eliminate the sphincter activity of the lower esophagus. Gastric acid can now reflux up into the esophagus, which has no defense against the corrosive acid and undergoes severe inflammation to the point sometimes of stricture. 'Heartburn' is a symptom of acid regurgitation into the esophagus. Surgical correction is often required and is successful if a competent lower esophageal sphincter is restored.

Stomach. Figure 3.3 shows the anatomical parts of the stomach. It is useful to know these as they show up well on x-ray, accuracy in lesion location is important for communication purposes and various parts have differing functions. The cardia, for instance, is the site of sphincter activity preventing destructive regurgitation of gastric acid into the esophagus; the body of the stomach contains the acid-secreting cells; the pyloric antrum is the motor center that coordinates gastric peristaltic emptying and is also the area that secretes acid-stimulating gastrin (a hormone; the antrum does not secrete acid); the pylorus is a sphincteric muscle controlling the expulsion of gastric content into the duodenum.

The lesser omentum (gastrohepatic ligament, a remnant of primitive ventral mesentery) stretches between the liver and lesser curve of the stomach. Its liver attachment

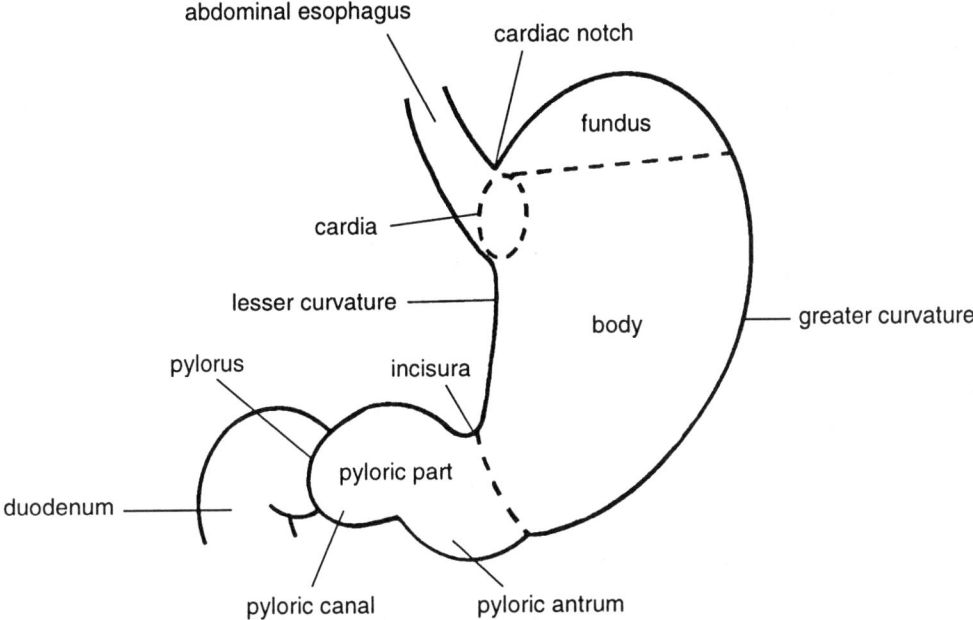

abdominal esophagus

cardiac notch

fundus

cardia

lesser curvature

greater curvature

pylorus

incisura

body

duodenum

pyloric part

pyloric canal

pyloric antrum

Figure 3.3   Stomach

is between the caudate and lateral segments of the left lobe. At the site of liver attachment is a groove running from liver hilum to inferior vena cava, which houses the ductus venosus in the fetus. The ductus venosus is the terminal portion of the umbilical vein which bypasses the liver *in utero*. Remember that the ventral mesentery continues laterally between the liver and proximal duodenum as the hepatoduodenal ligament, ending here in a free border. At its gastric lesser curvature attachment lie the left and right gastric arteries from the celiac and hepatic arteries. Corresponding veins and lymphatics accompany these lesser curve vessels. Also within the gastrohepatic ligament are the anterior and posterior vagus nerves, which distribute branches to the anterior and posterior aspects of the stomach before plunging into a hepatic nerve plexus in the region of the liver hilum (anterior vagus) and into the celiac plexus around the large visceral branches of the aorta (posterior vagus). (Sympathetic fibers reach the stomach from the celiac plexus by riding on the walls of the gastric arteries.) Remember that the twist of the stomach has changed the left and right vagi of the thorax into anterior and posterior vagi in the abdomen.

Duodenal ulcer is a common human affliction and is due in part to a hypersecretion of gastric acid. The vagi are largely responsible for this. Interruption of the vagus does indeed control acid production. Modern antacid vagal surgery attempts to interrupt only the acid-stimulating vagus fibers passing to the body of the stomach, while preserving normal gastric motility and emptying, and vagal innervation of the rest of the gut, by taking advantage of local vagal anatomy. Anterior and posterior vagal trunks enter with the esophagus. Interruption here will interrupt all vagal function, resulting in gastric atony and retention plus less well-defined functional deficits of the rest of the viscera. It is possible to interrupt only those fibers running to the acid-secreting stomach body as they branch off from the main trunks of the vagi. The bulk of the nerve can be preserved so that antral activity and proper emptying continues, and vagal innervation to the foregut organs and the remainder of the gut remains normal. Only gastric-acid secretion is diminished ('highly selective vagotomy').

Attached to the greater curvature side of the stomach is the old dorsal mesentery, now seen as the gastrosplenic omentum and the greater omentum, extending down over the transverse colon. At the attachment to the greater curvature lies the gastroepiploic arc, made up of right and left gastroepiploic arteries from the gastroduodenal and distal splenic arteries, respectively. Also from the distal splenic, short gastric arteries travel through the gastrosplenic omentum to the body and fundus of the stomach. The greater omentum extends downwards over the transverse colon to which it is attached and passes over the small intestine. It seems to function as the 'policeman' of the abdomen by walling off inflammation, stimulating the growth of new vessels where needed, and so on. Remember that this is a peculiar overgrowth of the primitive dorsal mesentery of the foregut. In its gastrocolic extent, together with the stomach and the lesser omentum, it forms the front wall of the lesser sac.

Duodenum (*duodeni* = L. twelve at a time).   This is composed of the initial 12 finger-breadths of the small intestine. It has four parts: superior, descending, transverse, and ascending portions. They form a large C around the head of the pancreas. Distal to the first part, the structure is 'extraperitoneal'. The third or transverse part is crossed by the superior mesenteric vessels. The ascending part crosses the aorta and rises to the duodenojejunal junction just to the left of the aorta. The hepatoduodenal ligament is attached in the adult to the first part. The ampulla of Vater is located approximately halfway along the descending portion on its posterior aspect. This is where the foregut budded off the precursors of ventral pancreas, biliary system and liver. Thus, the duodenum is foregut down to and just beyond the ampulla, and midgut thereafter. The foregut part receives blood from the artery of the foregut, the celiac, via the gastroduodenal artery and its superior pancreaticoduodenal branch; the midgut part is fed by the artery of the midgut, the superior mesenteric and its inferior pancreaticoduodenal branch. The two pancreaticoduodenal branches freely communicate and provide an important backup system in cases of injury to either the hepatic or superior mesenteric arteries. Within the duodenal C, and tightly abutting and sharing a blood supply with it, is the head of the pancreas. It is impossible to remove the pancreatic head without devascularizing the duodenum.

Remembering the gut rotation, it is then possible to understand the crossing of the superior mesenteric vessels ventral to the duodenum. On occasions, pressure from these vessels may obstruct the duodenum by squeezing it between the superior mesenteric artery and the posterior aorta. This results in gastroduodenal dilation and vomiting - the so-called SMA (superior mesenteric artery) syndrome. If you encounter a patient with these symptoms and signs, and a barium-swallow x-ray study shows the area of obstruction, you should be able to understand and diagnose the difficulty immediately!

Pancreas.   The 'sweetbreads' develop from buds into both dorsal and ventral mesenteries: that into the ventral mesentery is actually a bud from the foregut (duodenal) hepatic diverticulum, which also produces the biliary tract. Rotational forces swing the ventral pancreas around behind the duodenum to lie alongside the dorsal pancreas; the two then remain nestled within the tight curve of the C of the duodenum. The two anlagen fuse. The ventral pancreas forms the caudal pancreatic head and the uncinate (hooked) process. The uncinate is the lateral tip of the rotated ventral pancreas which extends behind (dorsal to) the superior mesenteric artery (in contrast to the ventral relation of the neck of the pancreas to the SMA). The dorsal pancreas forms the cranial portion of the head, and the neck, body and tail of the adult pancreas. The tail abuts the hilum of the spleen. Dorsal bud pancreas and spleen both develop in the dorsal mesentery, the spleen slightly more cranially. Remember that subsequent apposition of visceral dorsal mesenteric peritoneum with parietal posterior abdominal peritoneum – due to rotation of the stomach – ends with fusion and disappearance of the two touching peritoneal layers. Fusion renders the adult pancreas a retroperitoneal organ. It lies in the posterior wall of the lesser sac.

Not surprisingly, the two primordial pancreatic segments each have their own ducts leading to the duodenum (the site of original budding). The ventral pancreatic duct empties into the duodenum with the common bile duct. This pancreatic duct usually enters the medial wall of the terminal bile duct just proximal to the bile duct opening at the summit of the papilla of Vater. The dorsal pancreatic duct enters the duodenum 1–2 cm proximal to the ampullary opening. Frequently, with fusion of the two pancreatic segments, the two ducts communicate. The ventral duct and the proximal part of the dorsal duct (in neck-body-tail) form the main excretory duct (of Wirsung). The terminal part of the embryonic dorsal pancreatic duct remains small, retains its connection with the lumen of the duodenum and serves as the accessory pancreatic duct (of Santorini); its opening into the duodenum is often seen as the minor pancreatic papilla. Note that anatomical variations of adult duct anatomy are common, for example, failure of the two to communicate or the duct of Santorini is the main duct.

The pancreatic neck-body-tail are supplied by branches of the splenic artery and vein. The splenic trunks create a groove in the superior border of the organ.

Liver - Gallbladder - Biliary tract. Embryologically, these all originate as a bud or diverticulum from the distal foregut into the ventral mesentery. The orifice of this diverticulum is retained as the common duct orifice in the papilla of Vater (see below). The liver, rapidly developing in the ventral mesentery, soon grows upwards into the developing diaphragm - the septum transversum. The two structures lose their peritoneal coverings by fusion at the junction site. Thus, there is a posterior bare area of peritoneum-free liver (and diaphragm) in the adult.

Around the relatively circular margin of the bare area, the peritoneum of the old ventral mesentery is still attached to the liver to form the coronary ligament. This is readily discerned by following the falciform ligament to the liver. The right and left triangular ligaments of the liver are simply small wing-like extensions of ventral mesenteric peritoneum from right and left lateral aspects of the liver to the diaphragm. They are of no support value and of no clinical significance except for getting in the way of the surgeon during upper abdominal procedures. (They may be divided with impunity.)

The liver is held in position by the bare area attachment to the diaphragm and the support of the vena cava embedded in its posterior surface. The duodenal end of the hepatic bud forms the extrahepatic biliary ducts and the gallbladder. All of this growth occurs within the two peritoneal layers of the ventral mesentery. All structures to and from the liver thus must also travel within the leaves of this mesentery – in the adult, the hepatoduodenal ligament. More medially, this part of the ventral mesentery becomes the hepatogastric ligament or lesser omentum. As *bona fide* residents of the ventral mesentery, all of these structures are part of the anterior wall of the lesser sac. The liver with its bulk and bare area effectively blocks off entry into the sac save for the small opening between the posterior liver bare area and the retroperitoneal duodenum, the foramen of Winslow. In the anterior wall of the foramen – the hepatoduodenal ligament – run the hepatic artery, portal vein and bile duct (plus liver lymphatics and nerves). In posterior relation to the foramen is the parietal peritoneum-covered inferior vena cava.

The gross liver may be divided into anterior and superior diaphragmatic surfaces and a visceral surface (Figure 3.4). The anterior surface is smooth and peritoneum-covered, and touches, but is separate from, the parietal peritoneum-covered diaphragm. It is marked by the junction of the falciform ligament and liver. The peritoneal layers of the falciform spread over the liver dome and cranially form the coronary ligament margin of the bare area. Here the ventral mesentery peritoneum is reflected up onto the diaphragm. The posterior surface is largely the bare area encircled by the coronary ligament, and lying immediately under the pericardium and right atrium. The inferior vena cava issues here from its deep groove in the liver, receives the hepatic veins, then plunges through the diaphragm and pericardium to enter the right atrium.

Visceral surface. This is much more complicated! Marked by grooves making a large 'H', the right and left sagittal fissures form the sidebars of the H, and the liver hilum or porta hepatis is the crosspiece. The right sagittal fissure is formed by the gallbladder bed caudally, and the fossa or groove for the vena cava cranially. This fissure is interrupted in its midportion by a tongue of liver tissue, the caudate process. The left sagittal fissure is formed by the embryological umbilical vein (seen as the ligamentum teres in the free edge of the falciform ligament). The vein makes a notch where it reaches the sharp border of the liver from the falciform ligament. In its straight course from notch to inferior vena cava, it creases the underside of the liver. The ligamentum teres can be traced down in its crease to the left extremity of the porta hepatis. Here, in fetal life, it gives a small branch to the left portal vein, then continues across the liver to enter the inferior vena cava as the ductus venosus. The grooves of the ligamentum teres and ductus venosus form the left sagittal fissure. In the adult, it is difficult to discern any remnants of the ductus venosus, a large vital fetal structure.

The crosswise porta hepatis, the entrance to the liver, is around 4 cm long in the adult and houses the primary branching of the portal vein and hepatic artery, and the junction of the right and left hepatic bile ducts. Using these markers, and knowing the content of the porta, the liver may be divided into lobes: The right lobe is fed and drained by the right main branches of portal vein, hepatic artery and bile duct, and extends to the right from the right sagittal fissure; the left lobe comprises all the liver tissue lying to the left of the right sagittal fissure, and is divided into three sections by the porta and left sagittal fissure. Contained within the H are the caudate lobe cranially and quadrate lobe caudally. Beyond the left sagittal fissure is the lateral segment of the left lobe.

Advantage is taken of this anatomical lobar arrangement to effect safe and feasible segmental liver resection (*cf* lung segment resection). As in the lungs, the veins draining the liver segments end up as large vessels running between segments – hepatic veins – to the inferior vena cava. There are three major hepatic veins: the right hepatic vein

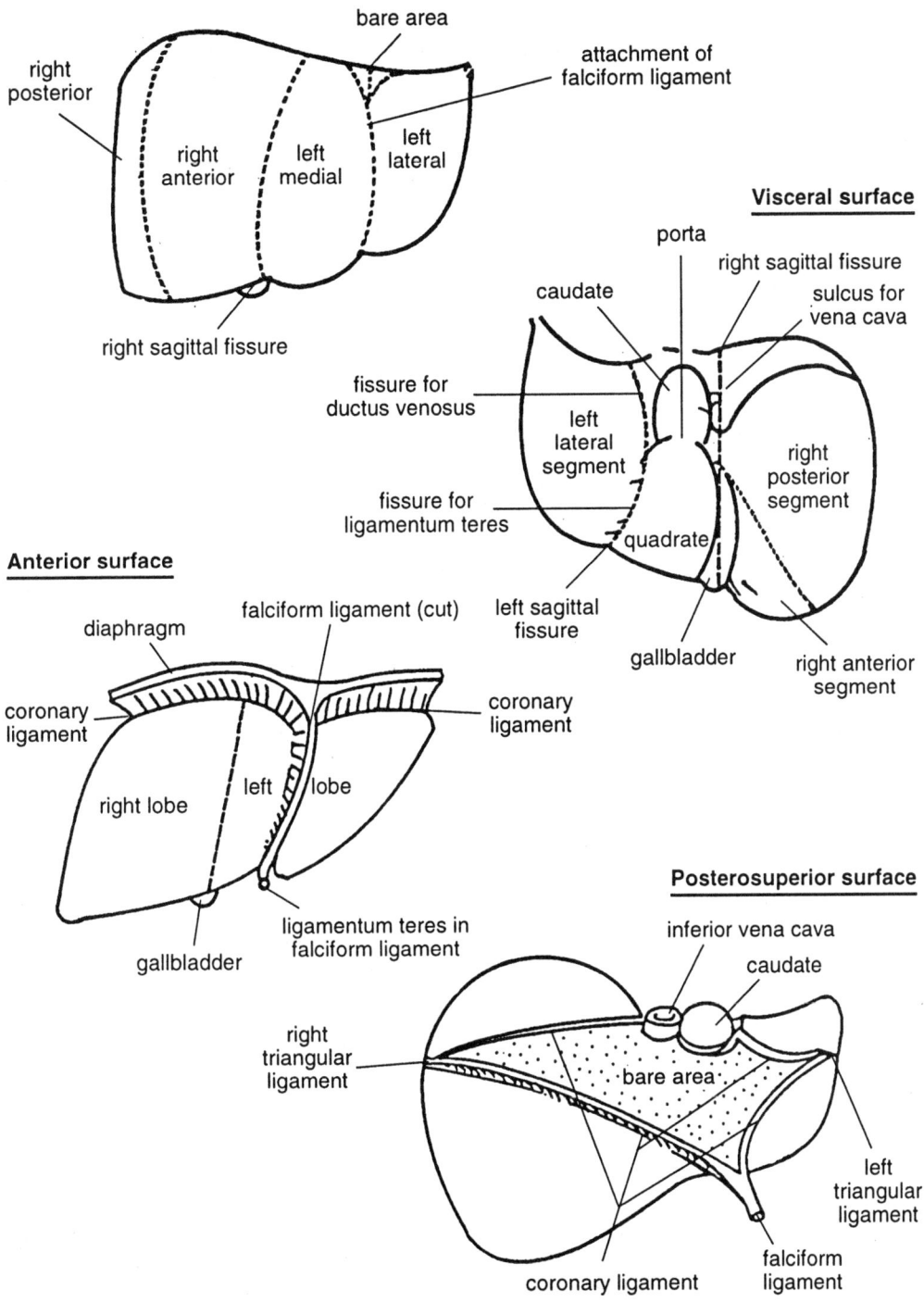

Figure 3.4 Liver segments

Visceral surface

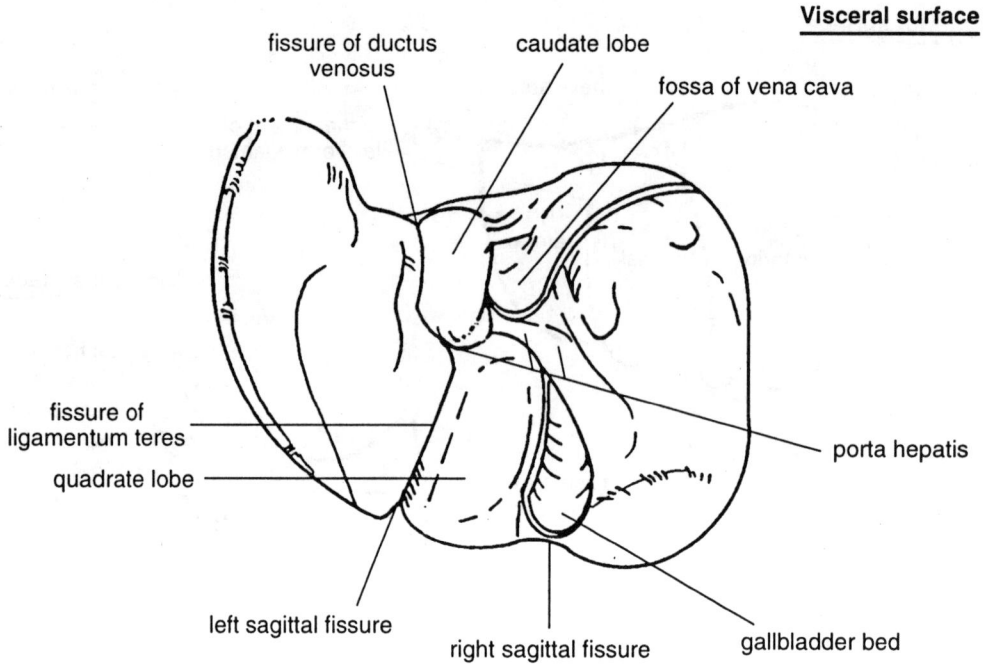

fissure of ductus venosus

caudate lobe

fossa of vena cava

fissure of ligamentum teres

quadrate lobe

porta hepatis

left sagittal fissure

right sagittal fissure

gallbladder bed

Visceral surface

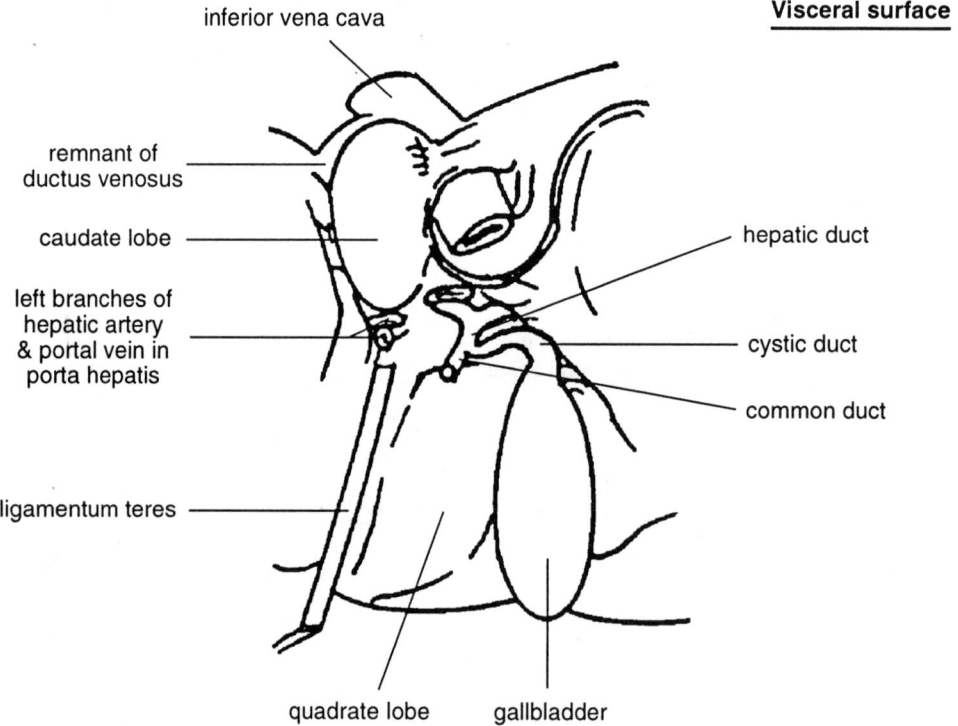

inferior vena cava

remnant of ductus venosus

caudate lobe

left branches of hepatic artery & portal vein in porta hepatis

hepatic duct

cystic duct

common duct

ligamentum teres

quadrate lobe

gallbladder

Figure 3.4 *continued*

runs between the anterior and posterior segments of the right lobe; the middle hepatic vein runs between the right and left lobes; and the left hepatic vein runs between the superior and inferior segments of the lateral segment of the left lobe. These veins are large, buried in liver substance and enter the vena cava at its diaphragmatic hiatus. During liver resection surgery, hepatic veins that drain the remaining liver tissue must be preserved or else that tissue will lose its drainage and not survive.

The visceral surface of the lateral segment of the left lobe is in contact with the abdominal esophagus and stomach whereas that of the right lobe is in contact with the duodenum, hepatic colic flexure and right kidney. These organs leave recognizable impressions on the visceral surface. The right lobar diaphragmatic surface is often grooved with impressions of overlying ribs.

Gallbladder and extrahepatic bile ducts. The right and left main hepatic ducts join within the liver hilum to form the common hepatic duct, which is housed between the peritoneal layers of the hepatoduodenal portion of the ventral mesentery (along with the portal vein, hepatic artery, and nerves and lymphatics running to and from the liver). At a variable distance from the hilum (1–4 cm), the common hepatic duct is joined by the gallbladder (cystic) duct. The two form the common bile duct which proceeds 7–8 cm to the duodenum, arching over the foramen of Winslow along the way. The common duct normally varies from 4–10 mm in diameter. Due to the foregut rotation, which twisted the duodenum as well as the stomach, the common duct in the adult lies posterior to the duodenum and pancreatic head running in a groove in the posterior aspect of the latter. In its terminal portion, the duct is surrounded by a controlling circular sphincter muscle (sphincter of Oddi) which makes a small swelling or papilla (papilla of Vater) in the posterior wall of the midportion of the duodenal C. At the apex of the papilla lies the opening of the duct. The main pancreatic duct usually empties into the papillary portion of the bile duct. Thus, the papilla represents the site of emergence of the hepatic diverticulum (bud) in the fetus, the bud that spawned the liver and its ducts, and the ventral pancreas.

The gallbladder has already been located in a relatively deep bed in the dependent portion of the right sagittal fissure covered by the peritoneum of the ventral mesentery. It is connected to the common hepatic duct by its cystic duct and supplied with arterial blood by a branch of the hepatic artery – the cystic artery – which usually arises from the right hepatic, but may arise from any or many branches of the hepatic arterial tree.

The anatomy of the extrahepatic bile ducts, gallbladder and cystic duct, and the blood supply of these structures and the liver, are extremely variable. The cystic duct may travel independently to the duodenum; the cystic artery may arise from any hepatic arterial branch; the right hepatic artery may be totally replaced or duplicated by an anomalous branch of the superior mesenteric artery, and a similar aberration may be found in the blood supply of the left lobe with branches from the left gastric artery. The foregut area is a frequent site of surgical disease. It contains vital organs and vessels. The usual anatomy and possible variations must be well known to every operator. The actual anatomy in every given case must be dissected out clearly before any definitive cuts or ligations are carried out.

Spleen. The spleen is not a digestive organ but is embryologically a close relation, as it develops in the dorsal mesentery along with the gut. We have already described how the dorsal mesentery swings over to the left when the stomach twists and how the dorsal mesentery from the spleen back to the midline is placed against the parietal peritoneum from midline laterally to kidney level, and how these two abutting surfaces fuse. After fusion, the remaining dorsal mesentery runs from the peritoneum over the kidney to the spleen as the splenorenal ligament.

The splenic artery courses from the celiac axis to the splenic hilum along the dorsum of the pancreas, giving branches to the pancreas as it goes. Just before entering the spleen, the artery gives off a left gastroepiploic branch, which joins a right gastroepiploic branch from the gastroduodenal artery. The two epiploics form an arc along the greater curvature of the stomach which is potentially fed from either end. Also, from its distal end, the splenic artery gives off short gastric arteries that run up between the layers of the gastrosplenic omental portion of the dorsal mesentery to the fundal area of the stomach. The splenic vein accompanies the artery to a point behind the neck of the pancreas where it joins the superior mesenteric vein to form the portal vein. Just before this junction, the splenic is joined by the inferior mesenteric vein which drains the left colon.

The spleen itself lies high and posteriorly in the abdomen between the stomach fundus and the diaphragm, generally in the scapular line. Its long midaxis lies along the level of the proximal ninth rib, and is also related posteriorly to the left adrenal and upper kidney. It is covered by the rib cage anteriorly and cannot be felt through the abdominal wall unless pathologically enlarged. It is composed of a homogeneous spongy material and varies greatly in size and weight, depending on its functional state, from 50–400 g. A normal adult spleen should fit readily into the cupped hand.

Midgut structures. The midgut includes the distal duodenum, jejunum, ileum and large intestine up to the splenic flexure area. Its artery is the superior mesenteric artery, which originates from the aorta behind the neck of the pancreas, at level L1, about 1.25 cm distal to the celiac artery. It passes forwards along with the superior mesenteric vein out from under the pancreatic neck, crosses ventrally over the uncinate process of the pancreas (tip of the rotated ventral pancreas), ventrally across the duodenum and out into the root of the old dorsal mesentery. The mesenteric root with the contained superior mesenteric artery extends diagonally from top left to bottom right of the abdomen, with a total length of around 15 cm. These 6 inches service around 20 feet of intestine, made possible by the ruffled fan-like structure of the developed mesentery. Vascular branches to the small intestine arise from the left lateral aspect of the artery (jejunal and ileal vessels) whereas those to the pancreas, ascending and transverse colons arise from the right lateral aspect (pancreaticoduodenal, midcolic, right colic and ileocolic vessels).

Small intestine. The junction of jejunum and ileum is not a fixed identifiable point. Typical jejunum and typical ileum, however, can be readily identified. The jejunum is of a larger diameter and of a thicker wall than the ileum. It is more active in the digestive process, contains more secreting glands and has an enhanced surface area due to multiple mucosal folds (valvulae conniventes). It is more richly vascularized than the ileum; in life, it is considerably pinker. Blood vessel architecture differs as well. In the jejunum, intestinal branches issue from the superior mesenteric and divide to join neighboring branches, and thus form an arcade from which long terminal branches (vasa recti) run to the jejunum. Ileal arteries divide 2–5 times, forming multiple small arcades. Terminal branches are short. A final difference is apparent in the disposition of fat in the small bowel mesentery. In the jejunum, fat is sparse as the mesentery approaches the gut. There are usually fat-free 'windows' in the mesentery at the gut margin. In the ileum, fat usually extends to and over the gut itself to a variable degree. Understanding these differences is an important aid to the abdominal surgeon in identifying his whereabouts in an obese adhesion-scarred abdomen!

The embryological omphalomesenteric duct is located about 2 feet from the ileocecal valve. If not normally totally resorbed, a diverticulm will be found in this terminal ileal area (Meckel's diverticulum). The diverticulum may cause clinical difficulties, especially in children, in the form of bleeding (from contained ectopic gastric acid-secreting mucosa) or by forming the leading point of an obstructing intussusception in which the intestine tries to pass the tic along as if it were a piece of meat.

The junction of the small and large intestines is marked by a readily palpable circular sphincteric ileocecal valve. It projects into the ascending colon lumen often far enough to be visible in barium x-rays of the large intestine (barium enema). Under normal conditions, the valve completely prevents reflux of colonic content into the ileum. (The resistance of the valve may be overcome under artificial conditions, such as the sudden increases in pressure exerted during barium enema examinations. The radiologist is usually able to visualize the terminal ileum.) Also, in the elderly, the valve is often incompetent.

Large intestine. When describing the whole of the large intestine, remember that the splenic flexure, descending and sigmoid portions are hindgut and supplied by the hindgut artery, the inferior mesenteric. For accuracy and clarity, realize that the large intestine extends from the ileocecal junction to the anus. Its parts are cecum (caudal to the ileocecal junction), colon (ascending, transverse, descending and sigmoid segments), rectum and anal canal. The post-rotational disposition of the large intestine has already been described.

The unique structural features of the large intestine down to the rectum depend on the anatomy of its muscle coats. The outer longitudinal layer is gathered into three distinct recognizable bands or teniae (except for the appendix). The inner circular layer is complete throughout. Between the teniae, characteristic ballooning sacculations of the less well-supported intestinal wall are found - the colonic haustra. In addition, small sacs of peritoneum-covered fat are found dangling from all parts of the intraperitoneal large intestine. These are up to 2–3 cm in length and are termed epiploic appendages.

The large intestine is, of course, considerably larger in diameter than the small intestine. On a plain x-ray film of the abdomen, the gas-filled large intestine may be recognized by its position, size and haustration. In the open abdomen, further identification is made by recognizing the presence of teniae and epiploic appendages on the colon, and the attachment of the greater omentum to the transverse colon.

The left colon and rectum are supplied by branches of the inferior mesenteric artery: the left colic artery supplies the splenic flexure and descending colon; the sigmoid arteries are of a variable number; and the terminal superior rectal branch supplies the rectum.

The large intestine serves as a fluid absorber and storage area for feces. It is not a vital organ as it may be bypassed surgically without clinical detriment, other than the risk of dehydration and electrolyte loss, by opening the ileum or colon to the abdominal wall. Contained material is allowed to empty into bags pasted to the abdominal skin (ileostomy, colostomy). Such operations are frequently done in cases of colonic cancer, diffuse ulcerative colitis, colonic obstruction, and so on.

The appendix vermiformis (cecal appendix) originates from the caudal base of the cecum well below the ileocecal valve. It is the familiar cause of much human illness and surgery. This vestigial finger-like structure (important in the digestive process of lower animals) of varying size extends outwards in any direction into the abdomen. It may be found as a pelvic structure relating to the ureter, bladder, uterus or rectum, or it may tuck into a retrocecal position or stray into a hernia sac. It may vary from 4 –15 cm in length and 4 –15 mm in diameter. It is supplied by its own appendiceal artery from the terminal superior mesenteric which runs in its own little appendiceal mesentery. The surgeon dealing with an inflamed appendix must be aware of all these possibilities. It must also be appreciated that, if the colon is incompletely rotated, the appendix will not lie in the right lower quadrant and must be sought by retracing the rotational route of the large intestine.

Rectum. This part of the large intestine, around 12 cm long, runs down to the anal canal from the upper sacrum level – around S3. It originates where the sigmoid straightens out, and is covered on its front and sides by peritoneum, thereby forming the posterior wall of the rectouterine (female) or rectovesical (male) pouches (cul-de-sacs). Descending and following the curve of the sacrum, the rectum soon becomes entirely retroperitoneal. As it reaches the level of the prostate, the rectum penetrates the levator ani diaphragm and is pulled forward by the puborectalis portion of the levator – the puborectalis sling. It then dives backwards and exits as the anal canal through the internal and external anal sphincter muscles to the anus proper. The anal canal is described as that segment running from the puborectalis sling to the anus or the segment from the dentate line of the rectum to the anus. (Further elucidation of this portion will be considered in the next section covering the pelvis and perineum.) Extending bilaterally from the distal extraperitoneal rectum are poorly defined fibrous structures termed the lateral ligaments of the rectum. They are of insignificant strength and provide little support. They serve as conduit areas for vessels, nerves and lymphatics to and from the rectum and lateral pelvic walls.

The major blood supply of the rectum is the superior rectal artery, the terminal branch of the inferior mesenteric. Its accompanying superior rectal vein drains up into the inferior mesenteric–portal vein system. The small middle rectal artery and vein are attached to the internal iliac artery and vein on each side. These unimportant vessels reach the rectum through its lateral ligaments. They are, however, of clinical interest as they offer a route for the spread of lower rectal disease to the walls of the pelvis, systemic venous system and lymph nodes rather than via the usual portal vein-mesenteric node-lymphatic route of the intestine to the liver.

Gut circulation and nerve supply.

Arterial supply. Three large anterior branches of the aorta are devoted to the GI tract: celiac (foregut); superior mesenteric (midgut); and inferior mesenteric (hindgut). Their branching with the various GI parts has already been described. It is important to know that all three are interconnected, thereby providing a collateral circulation if any part is occluded. The pancreaticoduodenal arcades connect celiac and superior mesenteric arteries. The common hepatic artery may be tied off with impunity proximal to a patent gastroduodenal artery. The superior and inferior mesenterics are connected by branches of the superior mesenteric midcolic and the inferior mesenteric left colic artery. These branches run in the mesentery close to the border of the colon, forming the marginal artery (of Drummond). A sudden occlusion of the superior mesenteric will inevitably result in necrosis of the small intestine and proximal colon, and must be relieved – in emergency mode! A slow, gradual reduction in blood flow, however, will allow time for the collateral channels to enlarge while preserving the vitality of all areas. Numerous other small connections may be present; the above are the important constants. It is important to know the anatomy of the gut circulation.

The following outline includes the important main branches.

Celiac axis
    Splenic
        Pancreatic rami
        Gastroepiploic (left)
        Short gastrics
    Left gastric
    Hepatic
        Gastroduodenal
            Superior pancreaticoduodenal
            Gastroepiploic (right)
        Right gastric
        Right and left hepatics

Superior mesenteric
    Inferior pancreaticoduodenal
    Jejunal and ileal branches
    Iliocolic
        Appendiceal
    Right colic
    Midcolic

Inferior mesenteric
    Left colic
    Sigmoidal
    Superior rectal

(Marginal artery of Drummond)

Middle rectal (from hypogastric)
Inferior rectal (from internal pudendal)

Portal venous system. Virtually all of the above arterial blood passes through capillary beds in the gut and spleen, and is gathered up in the portal venous system and delivered to a second capillary bed in the liver. The portal vein is formed by the junction of splenic and superior mesenteric veins behind the neck of the pancreas and in front of the vena cava – level L2. It is around 8 cm in length. The inferior mesenteric vein joins the party, usually as a tributary of the distal splenic, just before the portal vein itself. The left gastric (or coronary) vein is a large tributary of the proximal portal vein. Withn the hepatoduodenal ligament during its course to the liver hilum, the portal vein lies posterior to the bile duct and hepatic artery and, of course, in the anterior boundary of the foramen of Winslow. Portal blood is distributed to the tissues of the liver in close company with branches of the hepatic artery and the bile duct. The combined portal and hepatic (arterial) blood passes through the hepatic capillary bed, and is then collected in the hepatic veins and finally returned to the heart via the inferior vena cava. The portal vein delivers around 70% of the volume and approximately 50% of the oxygen to the liver. The hepatic artery supplies half of the oxygen and most of the pressure. Hepatic artery mean pressure is normally around 100 mmHg whereas that of the portal vein is approximately 7 mmHg.

It is important to know the collateral routes taken by portal venous blood to reach the right heart if there is obstruction to passage of portal blood through the liver or hepatic veins. Obstructing liver disease, or thrombosis of the portal or hepatic veins is a frequent cause. Major connections with the much lower-pressure systemic circulation occur in three areas: between the left and short gastrics, and the azygos system in the gastroesophageal junctional area; between the superior rectal and inferior rectal of the internal pudendal-hypogastric system at the anorectal junctional area; and between veins in the falciform or reconstituted umbilical vein in the ligamentum teres and umbilical skin veins at the umbilicus.

With portal obstruction, portal pressure rises to 4–5 times above its normal value (portal hypertension). The above-mentioned collaterals are forced to transmit an unusual volume and pressure of blood. The result is the formation of bulging thin-walled esophageal varices, enlarged anorectal veins (hemorrhoids) and coils of large veins around the umbilicus (caput medusae). The spleen is uniformly enlarged by the increasing back-pressure. The significant clinical impact of all this, aside from the underlying problem in the liver (or its incoming or outgoing veins) and the discomfort of anal hemorrhoids, is the inability of the esophageal varices to withstand the stress. Thus, they frequently rupture with ensuing torrential hemorrhage into the esophagus and stomach. Measures are available to control the ruptured variceal bleeding (local esophageal/gastric balloon pressure via a swallowed tube; injection of thrombosing chemicals into the varices via esophagoscope and long needle). Similarly, measures to lower portal pressure (posterior pituitary hormone vasodilating action, surgical (portacaval) shunts between high-pressure portal and low-pressure systemic veins, or endohepatic radiologically controlled and placed conduits between portal and hepatic vein branches) are often successful. Portal hypertension is always a difficult and dangerous clinical problem.

Gastrointestinal lymphatics.   GI lymph channels follow the course of the GI arteries to the region of their posterior midline origins. Passing around the aorta, they enter a thin-walled lymphatic sac, the cisterna chyli. The cistern rests behind the aorta, and in front of the right diaphragmatic crus and body of L2. Interposed along the drainage route in the GI mesentery are innumerable lymph nodes which become obviously enlarged when a disease process involves the GI area drained, or when the nodes themselves become diseased. The lymph nodes are named for the artery they accompany, or the organ or region drained, for example, mesocolic nodes, inferior mesenteric nodes and pancreatic nodes. The cisterna extends upwards through the aortic hiatus of the diaphragm to become the thoracic duct. Digested fat is absorbed through GI lymphatic channels rather than through the portal system. Fat-rich GI lymphatics are seen as milky white vessels and are aptly termed 'lacteals' (L. milk).

Autonomics of the GI tract. Sympathetic preganglionics arise from the white rami communicantes of T5–12 and L1–2. They proceed to and through the sympathetic trunk without synapse, and continue on as splanchnic nerves to postganglionic sympathetic neuron cell masses clustered around the main visceral branches of the aorta (celiac, renal, superior mesenteric, inferior mesenteric ganglia). Postganglionic axons travel to all parts of the GI tract on the backs of the GI arteries.

Parasympathetic preganglionics are contained in the two vagi and pelvic splanchnic branches of S2–4. These preganglionics travel to all parts of the GI tract via GI arteries and synapse in parasympathetic ganglion cells in the visceral walls. Short postganglionics then give parasympathetic innervation to visceral smooth muscle and glands.

Large plexuses of nerve fibers – both sympathetic pre- and postganglionics, and parasympathetic preganglionics – plus masses of sympathetic postganglionic neurons are found around the abdominal aorta and its visceral artery branches. Nerve fibers passing to the viscera endow these vessels with a tough collar as they ride the arteries to their destinations.

Visceral sensory afferents that record pain, and are triggered by distention or spasm, travel to the spinal cord by retracing the route taken by the sympathetic efferents to the viscera. In the cord, their message is carried to consciousness as is any other sensory signal. Afferents recording satiety, chemical status, pH, etc., are thought to reach the central nervous system via the vagi and S2–4 roots. Parasympathetic afferents are largely subconscious.

Abdominal retroperitoneum. Description of this area allows us to pick up loose ends, and generally tidy up and package the abdominal vascular, neural and lymphatic anatomy. The abdominal aorta is easily displayed. It ends by bifurcating into the common iliac arteries in front of the body of L4. Note that this is just below navel level and well above the palpable sacral promontory at L5–S1. Thus, the take-off point of its inferior mesenteric branch is 4 cm above the bifurcation and, thus, a good 2 cm above the navel – level L3. All of this is located higher than you might think! The urinary tract, suprarenal glands, retroperitoneal muscles and lumbar plexus are included.

Retroperitoneal arteries.

Abdominal aortic branches.
Parietal
    Inferior phrenic
    Lumbar segmental
    Middle sacral

Terminal
    Common iliac

Visceral
    (Celiac)
    (Superior mesenteric)
    (Inferior mesenteric)
    Middle suprarenal
    Renal
    Testicular/ovarian

Inferior phrenics. These arise from the sides of the aorta just below the aortic hiatus. They help supply the diaphragm and give off superior suprarenal branches.

Middle suprarenals. These arise from the sides of the aorta at superior mesenteric artery level and constitute the aortic contribution to the suprarenal supply.

Renals. These arise from the sides of the aorta just below superior mesenteric artery level. On the right, the artery passes behind inferior vena cava, renal vein, head of pancreas and duodenum to reach the kidney. On the left, the artery passes behind the renal vein and pancreas. Both give off inferior suprarenal and ureteral branches.

Gonadals. These arise from the front of the aorta just caudal to the renals.

Lumbars. Segmental vessels (*cf* intercostals), these pass deep to the diaphragmatic crura, psoas and quadratus lumborum. In addition to supplying tissues of the back and abdominal wall, they supply important radicular branches to the nerve roots and spinal cord through branches passing through intervertebral foramina.

Middle sacral. This small vessel from the back of the aorta descends in the midline over L4–5 and the midline sacrum. This, too, produces radicular branches to nourish nerve roots.

Retroperitoneal veins. The inferior vena cava has tributaries similar to the retroperitoneal branches of the aorta: inferior phrenics, suprarenals, gonadals and segmental lumbars. However, the gonadals and suprarenals on the left empty into the left renal vein (remnant of the embryological left subcardinal vein). (The left inferior phrenic empties into both the left renal and inferior vena cava.) In addition, the inferior vena cava receives the large hepatic veins at the diaphragmatic level. Connections between lumbar segmentals form ascending lumbar veins which, in turn, form the beginnings of the azygos and hemiazygos systems of the thorax.

Urinary tract. This totally retroperitoneal system originates outside of the celomic lining.

Kidneys. Lying on either side of the vertebral column, these organs extend from T12–L3. They are surrounded by considerable perirenal fat, the whole contained in an envelope of fascia (Gerota's fascia) branching from the retroserous fascia that steadies the entire peritoneum. Outside of the renal fascial envelope is another collection of fat – pararenal fat – lying between Gerota's fascia and the deeper transversalis fascia. Posteriorly, the upper kidney rests on the twelfth rib and the diaphragm. Thus, it is in important relation to the pleural costophrenic recess. The lower part of the kidney rests on psoas, quadratus lumborum and the posterior rib origin of the transversus abdominis muscle. Superiorly, the kidneys are capped by the suprarenal glands. Anteriorly, the right kidney is covered by the right lobe of the liver whereas the left is covered by the spleen. Both are related to colonic flexures – hepatic and splenic, respectively. Anteromedial relations are the duodenum on the right and the tail of the pancreas on the left.

Examination of the kidney reveals a strippable capsule, a cortex and a medulla; the latter consists of a series of cone-shaped pyramids projecting as a series of renal papillae into a urine-collecting pelvis. The pelvis embraces the papillae in cup-shaped renal calyces (Gk. cups). The pelvis tapers down to a smooth-muscled ureter capable of milking urine by peristaltic action down to the bladder.

The blood supply to the kidneys emanates from large lateral branches of the abdominal aorta. Starting just caudal to the superior mesenteric artery take-off – level of L1–2 disk – the right renal artery passes behind the vena cava, right renal vein, head of the pancreas and descending duodenum. Posteriorly, it crosses the crus of the diaphragm. The left renal runs over the left crus deep to the left renal vein, pancreatic body and splenic vein. The right renal vein runs directly medially into the lateral aspect of the vena cava. The left vein is longer as it has to run in front of the aorta to reach the inferior vena cava. During its course, it receives, from below, the left gonadal vein and, from above, the left suprarenal and inferior phrenic veins. On the right, these venous branches run directly into the vena cava. Remember that this part of the vena cava is formed when the two embryological cardinal veins develop connections and then decide to form one right-sided channel rather than a bilateral symmetrical system (*cf* the common iliac veins, the brachiocephalic veins, the vitelline veins). The remnant of the left subcardinal vein (now part of the left renal vein running to a single inferior vena cava) still receives the original tributaries of the primordial symmetrical bilateral cardinal system – the left gonadal, left inferior phrenic and left suprarenal.

Lymphatics from the kidneys follow the renal vessels medially to enter lumbar para-aortic node groups and, from there, to the cisterna chyli and thoracic duct.

Ureter.   Knowledge of the anatomy of this vitally important structure is imperative! It is at risk during any intra-abdominal surgical procedure. It is frequently the source of clinical signs and symptoms caused by neighboring organ disease. It is subject to many congenital anatomical anomalies (for example, duplication, ectopy, abnormal formation due to malformed or ectopic kidneys). The integrity of the ureter is essential for the health of the kidney it drains and, therefore, the health of the entire organism. Its injury (through inadvertent ligation or division resulting in kidney tissue destruction, extravasation of urine or uremia) is the quintessential surgical error!

In the abdomen, the normal right ureter issues from the kidney pelvis behind the descending part of the duodenum. Both ureters lie in subperitoneal areolar tissue resting on the lateral surface of the psoas. With mobilization of the ascending or descending colon by opening of the embryological fusion planes, the ureter usually clings to and comes up with the elevated mesentery of the colon. The ureter is at risk if this tendency is not recognized. The ureters slowly veer across the anterior aspect of the psoas from lateral to medial in their caudal descent. They enter the pelvis immediately anterior to the bifurcation of the common iliac arteries. This offers an easily palpable, pulsating landmark for locating the ureters. In its abdominal course, the ureter is crossed anteriorly by the gonadal vessels (these are also elevated with fusion-plane mobilization of the colon). The pelvic course of the ureter will be reviewed in the following section on the pelvis.

The arterial supply to the ureters comes from both above (ureteric branches from the renals) and below (branches from the hypogastric vesical arteries). There is therefore a potentially ischemic area in the midabdominal ureter which may become overt with excessive surgical ureteral mobilization, producing local necrosis, urinary leak or scarring stricture.

Nerves to the ureteric smooth muscle come from both segments of the autonomic system via the preaortic ganglia and plexuses, and the pelvic plexus. They travel on the ureteric vessels.

Afferent nerves conveying pain sensation for all of the viscera retrace the sympathetic path to the spinal cord and reach consciousness as do pain receptors anywhere else in the body. Kidney pain is thought to be due to swelling pressure on its indistensible capsule and is felt in the lumbar costovertebral angle (CVA). Ureteric pain is, as with gut pain, triggered by distention or spasm. As with any other hollow viscus, ureteric pain is of a colicky griping nature. It is usually felt anywhere around the loin from CVA to testis (remember the close anatomical embryological relationship between the gonadal and urinary systems).

Suprarenals. These are two firm 3–5 cm glands astride each kidney. They incline towards the celiac artery and its plexus so that the medial aspects of the two are only 5 cm apart. They are composed of a central medulla, which secretes epinephrine on preganglionic sympathetic stimulation, and an outer cortex, which secretes corticosteroid hormones on anterior pituitary hormone stimulation. The cortex is readily recognizable on section by its golden-orange color. Posteriorly the adrenals rest on the diaphragm. The right adrenal is related anteriorly to the edge of the bare area of the liver. It is covered by peritoneum reflected from both the posterior aspect of the coronary ligament and lateral aspect of the duodenum. The left adrenal is covered by peritoneum of the floor of the lesser sac, which separates it from the cardia of the stomach. More caudally, it is related anteriorly to the tail of the pancreas and the splenic vessels.

Arterial blood supply is profuse, from multiple branches of the aorta, inferior phrenic and renal arteries. The right suprarenal vein drains directly into the vena cava, and the left into the left renal vein (the 'left vena cava'). Nerves are supplied in force by the adjacent celiac and renal ganglia-plexuses.

The association of suprarenals with the kidneys is embryologically incidental. They are not 'blood relatives'.

Muscles of the retroperitoneum. The psoas major muscle is the familiar long rounded muscle running from the lumbar transverse processes, disks and bodies to the lesser trochanter of the femur. It flexes the thigh and the spine. One side acting alone bends the lumbar spine laterally. Psoas major is an interesting muscle because of its relations: The ureter and gonadal vessels cross it; the genitofemoral nerve pierces it and runs down its anterior midline; the lumbar plexus lies behind it, with the (important) femoral and obturator nerves hiding beneath its lateral and medial borders, respectively. Similarly, lumbar segmental vessels pass behind the psoas. The straight lateral border of the muscle is readily recognizable on abdominal x-rays. Absence or distortion of the psoas shadow indicates trouble in the posterior abdomen (infection, fluid collection, tumor, etc.).

Quadratus lumborum is the lateral neighbor of psoas. It attaches to the twelfth rib and to the tips of the lumbar transverse processes and the iliac crest. The muscle is a lateral spine flexor and, in addition, holds the lower ribs down during forced expiration.

Lumbar plexus. This is composed of the ventral primary rami of L1–4. Three very important nerves arise from the lumbar segmental nerves. The femoral nerve arises from L2–4 and courses downwards under cover of the lateral border of psoas, then under the inguinal ligament directly into the femoral triangle, where it arborizes to innervate the anterior thigh muscles. The obturator nerve, also arising from L2–4, runs downwards into the pelvis beneath the medial border of psoas, under the iliac vessels and along the wall of the true pelvis, and through the obturator foramen in the internal obturator muscle to the medial thigh, where it innervates the adductor muscles. Finally, the lumbosacral trunk, L4–5, constitutes a major contributor to the sacral plexus, the sciatic nerve. The entire output of L5 enters the sacral plexus whereas most of L4 contributes to the obturator and femoral nerves. The trunk is deep and passes over the ala of the sacrum to join the sacral plexus.

The L1 segment courses around the abdomen to the suprapubic skin. Lower segmentals from the plexus end in cutaneous areas of the lateral and anterior upper thigh. The ilioinguinal nerve – L1 – ends in an anterior cutaneous branch that runs along with the spermatic cord through the inguinal canal, exits through the subcutaneous inguinal ring, and supplies skin of the upper inner thigh and the abutting scrotal (labium majoral) skin. The nerve is at risk in any inguinal hernia operation. If injured, the resulting numbness may be a serious annoyance. The lateral femoral cutaneous branch of the plexus – L2–3 – runs under the inguinal ligament at the anterior superior iliac spine to innervate the skin of the lateral thigh. Impingement on the nerve at the anterior superior spine region causes a painful neuralgia in its lateral thigh area. The genitofemoral nerve – L1–2 – pierces psoas and runs down the anterior surface of the muscle. The genital branch innervates the cremaster muscle whereas the femoral branch supplies a small patch of thigh skin just below the inguinal ligament.

Autonomics. GI features have already been described. The sympathetic trunks continue down into the pelvis just posterolateral to the inferior vena cava and the aorta. Ganglia are present in varying numbers and supply postganglionic sympathetic fibers to the segmental nerves via the gray rami communicantes. (Preganglionic white rami communicantes are present only as far down as the L2 level; distal to L2, preganglionic fibers must travel down the trunk to reach and synapse within the appropriate sympathetic ganglion. The parasympathetic system has nothing to do with spinal nerve autonomic activities such as sweat and sudoriferous glands and vascular smooth muscle.)

Diaphragm.
The diaphragm is only around 4-mm thick and covered with a serous membrane on both sides (pleura and peritoneum). The whole of the central portion is innervated by the phrenic nerve. The rim of the diaphragm is derived from chest wall material and therefore innervated by intercostal nerves. Thus, sensory stimuli from either the abdominal or chest side of the **rim** are accurately localized. Pain from the central phrenic innervated part is referred to the top of the shoulder (phrenic nerve, C3–5).

Embryologically speaking, the diaphragm originates from the septum transversum (central tendon and muscle), pleuroperitoneal membrane, dorsal mesentery of the esophagus and the chest wall (Figure 3.5).

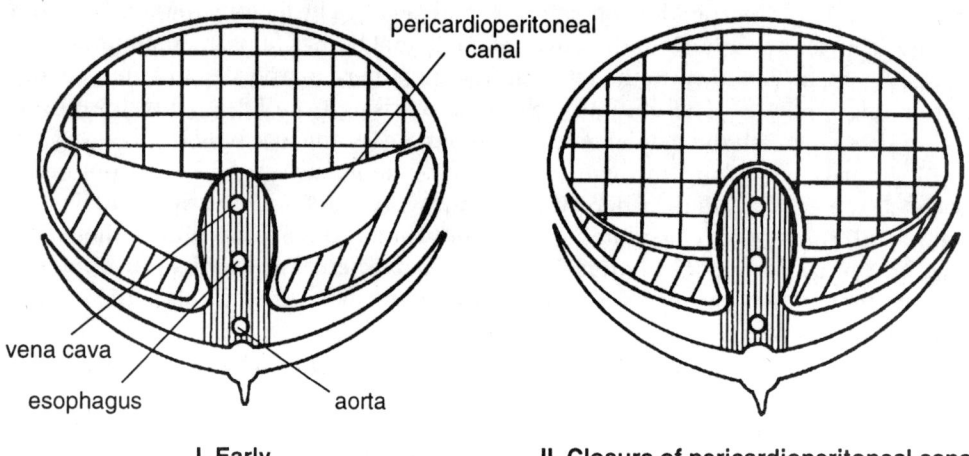

I. Early

II. Closure of pericardioperitoneal canal

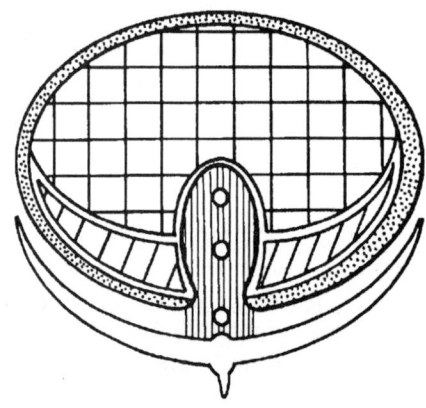

III. Excavasation of body wall

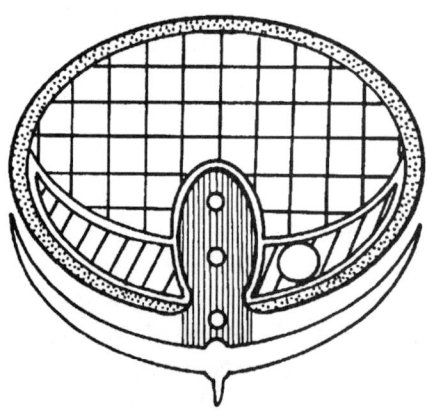

IV. Closure defect in pleuroperitoneal membrane (congenital diaphragmatic hernia)

Septum transversum

Pleuroperitoneal membrane

Esophageal mesentery

Body wall

Figure 3.5 Development of the diaphragm

Congenital diaphragmatic hernias are the result of defects in diaphragm formation. Incomplete development of the muscle adjacent to the xiphoid (where superior epigastric vessels normally pass) results in hernias which may become large enough to contain the entire stomach (Morgani hernia or 'upside-down stomach'). A more common congenital defect occurs when the pleuroperitoneal membranes fail to close the pericardioperitoneal canals completely (congenital diaphragmatic hernia of Bochdalek. The term 'pericardioperitoneal' is retained, although the canal at this stage connects only pleural and peritoneal areas.) This defect allows free access of the abdominal intestinal contents into the pleural cavity, causing compression hypoplasia of the ipsilateral lung, and distortion of the pleural cavities and mediastinum. Serious respiratory deficiency results and requires emergency correction – possibly even corrective intrauterine surgery.

Acquired diaphragmatic hernias are common findings, particularly the esophageal hiatus hernia, wherein the stomach slides up into the chest mediastinum through the esophageal diaphragmatic opening (see below). Another is the post-traumatic hernia, due to rupture of the diaphragm in response to a severe compressive force against the abdomen. Intestine then enters the chest and may cause respiratory embarrassment or becomes stuck, twisted or obstructed.

Abdominal aspect of the diaphragm.
Crura. These are two tendinous 'legs' extending from the bodies of L1–3. They rise and meet in an arch over the aorta at T12 level (aortic hiatus) to form a 'median arcuate ligament'. The thoracic duct and azygos vein accompany the aorta through this opening. (Thoracic splanchnic nerves pass through the structure of the crura.) From the right crus, a poorly defined band descends to attach to the duodenojejunal junction, the ligament of Treitz, supposedly to hold and support the junctional area. However, this has not proved to be a clinically significant entity.

Esophageal hiatus. This is the diaphragmatic muscle-encircled opening at T10 level that transmits the esophagus, vagi and veins connecting portal and systemic (azygos) systems (see discussion on portal hypertension and esophageal varices above). The stomach often slides through this hiatus to produce the familiar esophageal hiatus hernia, which interferes with the normal pressure relationships between the chest and abdomen involving the esophagus and stomach. Negative intrathoracic inspiratory pressure interferes with normal lower esophageal sphincter action. The lower esophageal sphincter normally resides in the abdomen but, when herniated into the negative-pressure thorax, sphincter action becomes incomplete and thus allows gastric acid to enter the 'acid-defenceles' esophagus. All sorts of inflammation results, producing symptoms ranging from acid heartburn to ulceration, and stricture of the esophagus. Many hiatus hernia cases require surgical correction.

Vena caval diaphragmatic hiatus presents as an opening at T8 level for the inferior vena cava lying immediately caudal to the pericardium and right atrium.

Lumbocostal arches. Posteriorly, the diaphragm springs from two tendinous arches over the psoas and quadratus lumborum muscles on each side: The medial arch passes over psoas from the crus to the transverse process of L1; the lateral arch springs from the transverse process of L1 and arches over quadratus lumborum to the twelfth rib. These are the medial and lateral arcuate ligaments, respectively.

Review Exercises

Frontal section kidney. Sketch in pelvis, calyces, cortex, columns, papillae and pyramids.

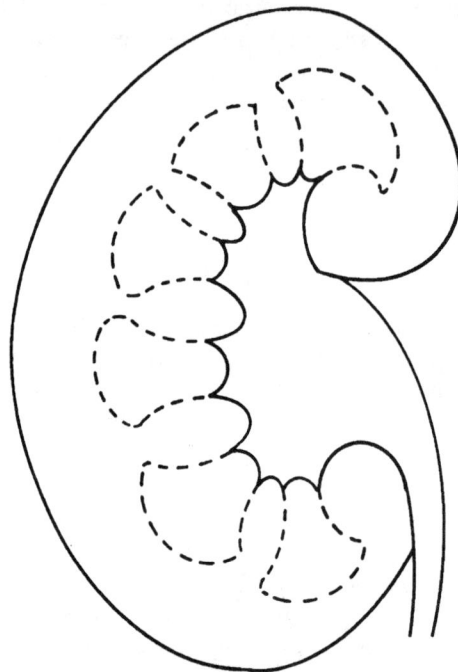

Sketch in branches of the abdominal aorta.

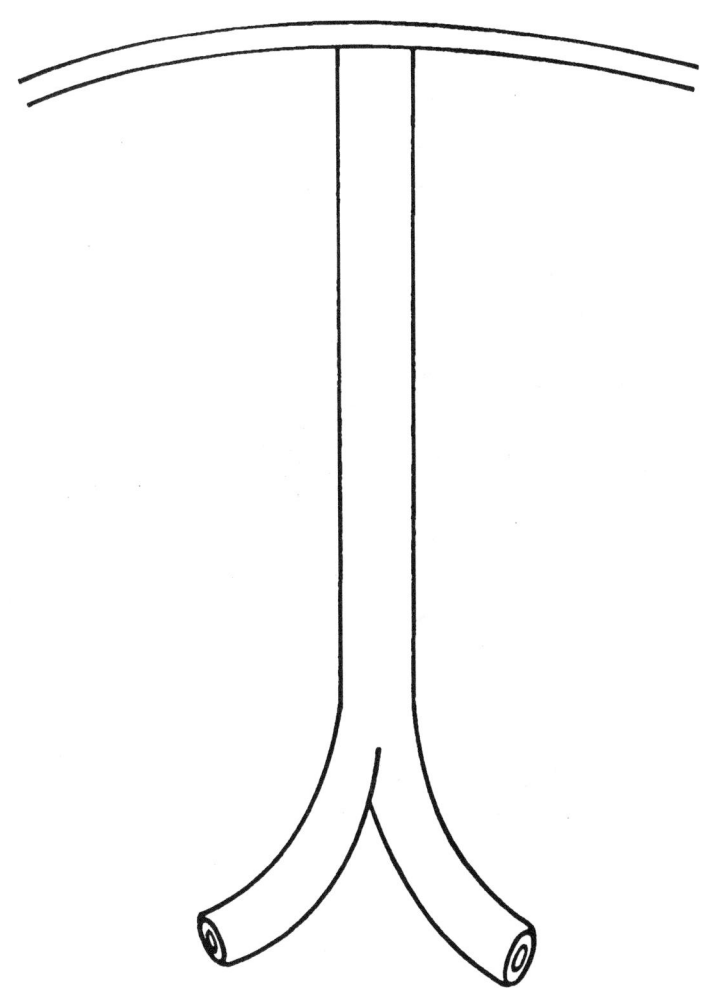

Sketch in branches of the celiac trunk.

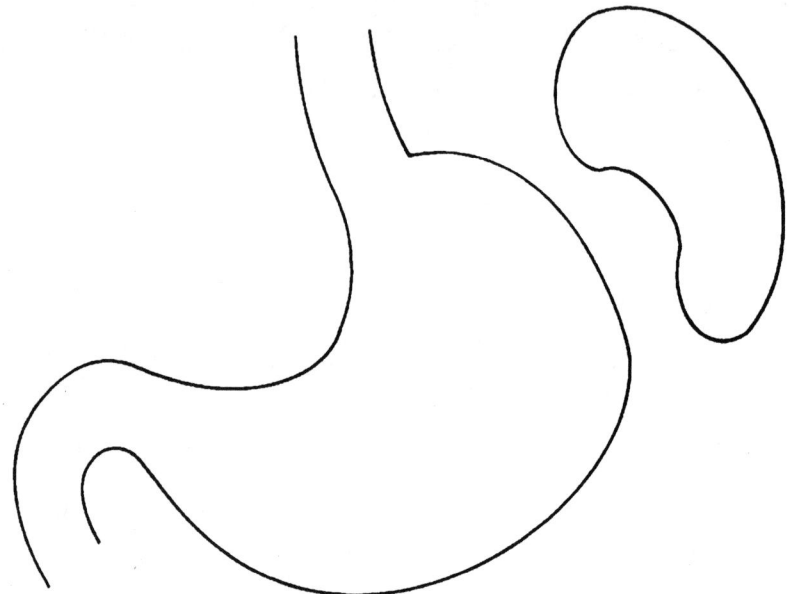

Sketch in branches of superior and inferior mesenteric arteries.

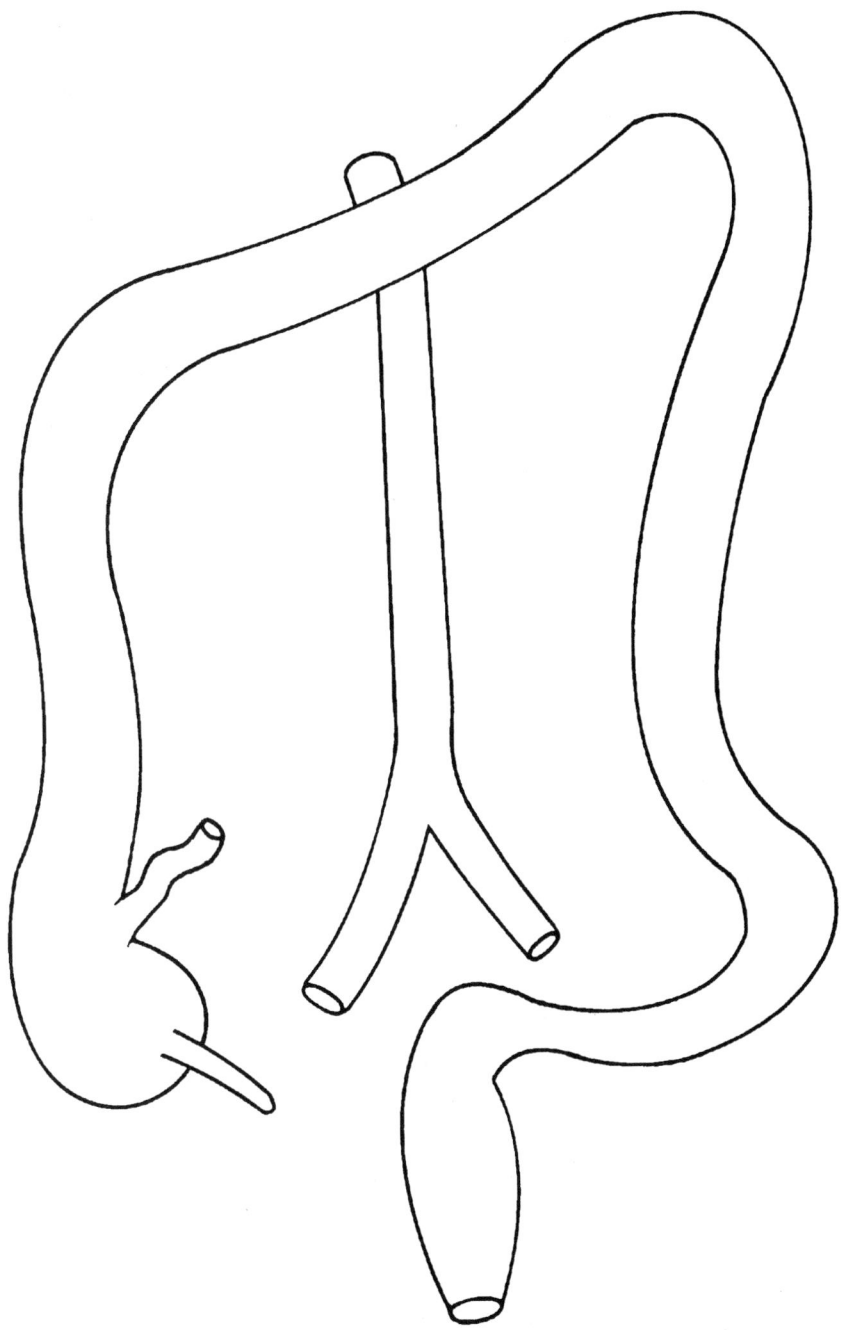

Sketch in branches of the lumbar plexus.

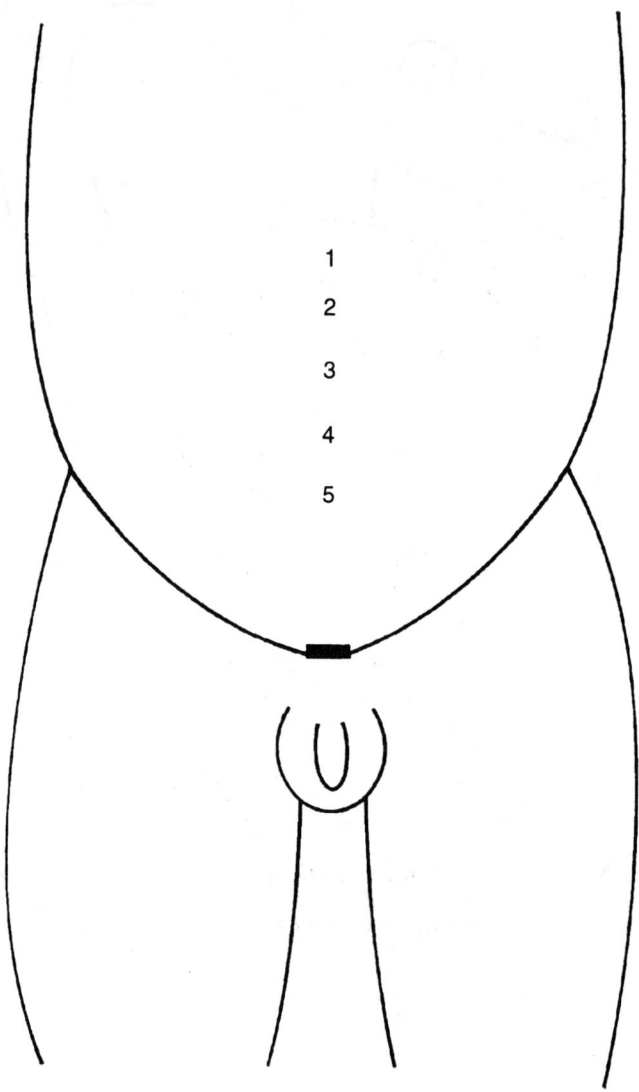

1
2
3
4
5

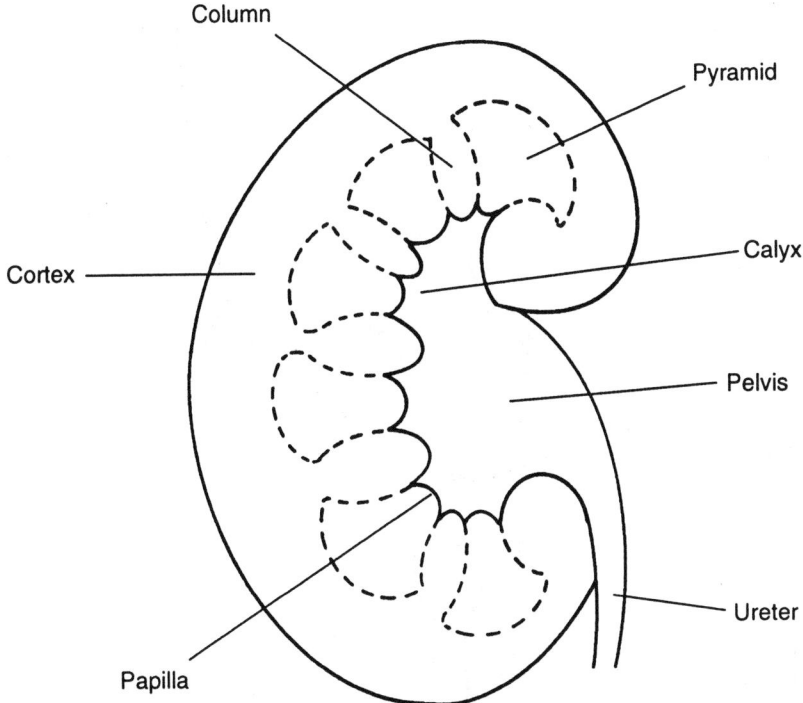

Column

Pyramid

Cortex

Calyx

Pelvis

Papilla

Ureter

Kidney: Frontal section

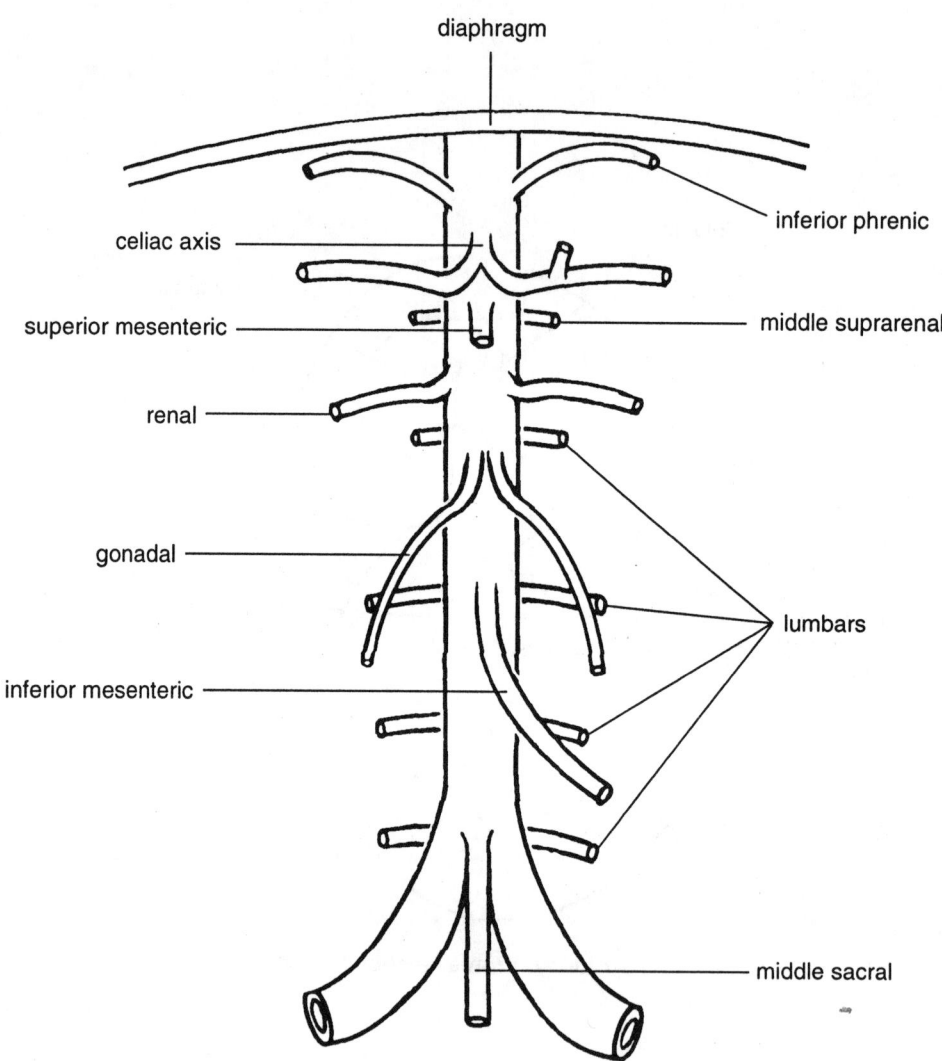

Abdominal aorta branches

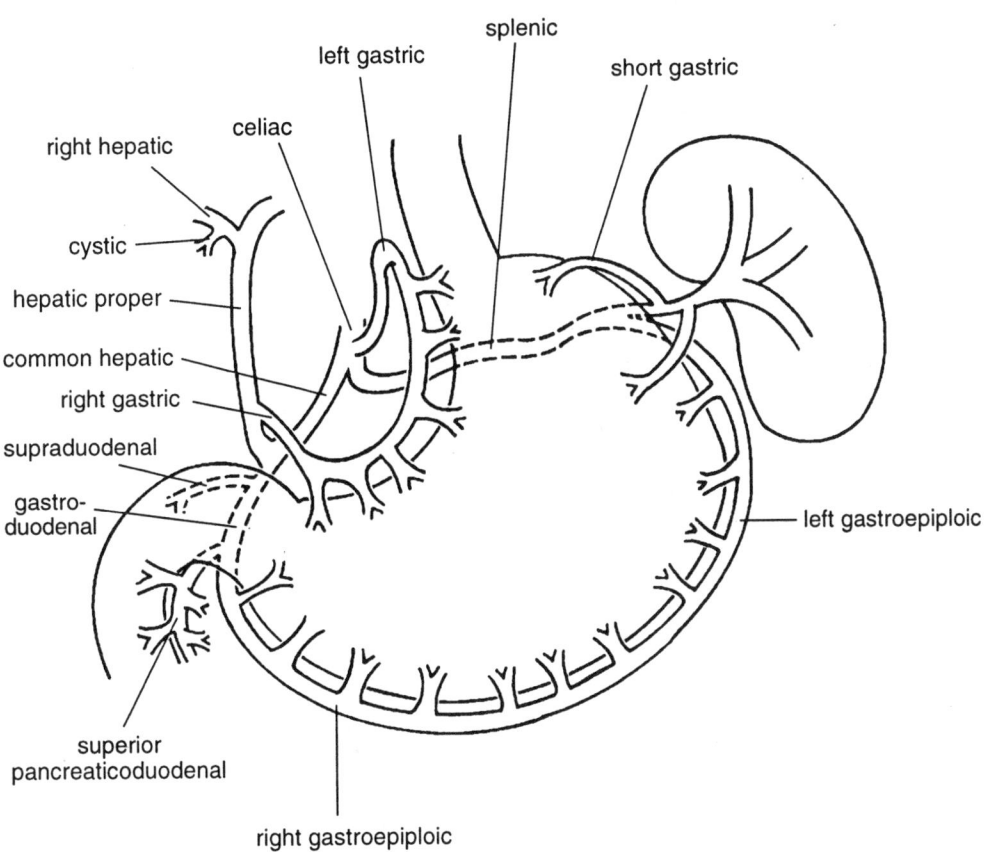

right hepatic

celiac

left gastric

splenic

short gastric

cystic

hepatic proper

common hepatic

right gastric

supraduodenal

gastro-
duodenal

left gastroepiploic

superior
pancreaticoduodenal

right gastroepiploic

Celiac trunk

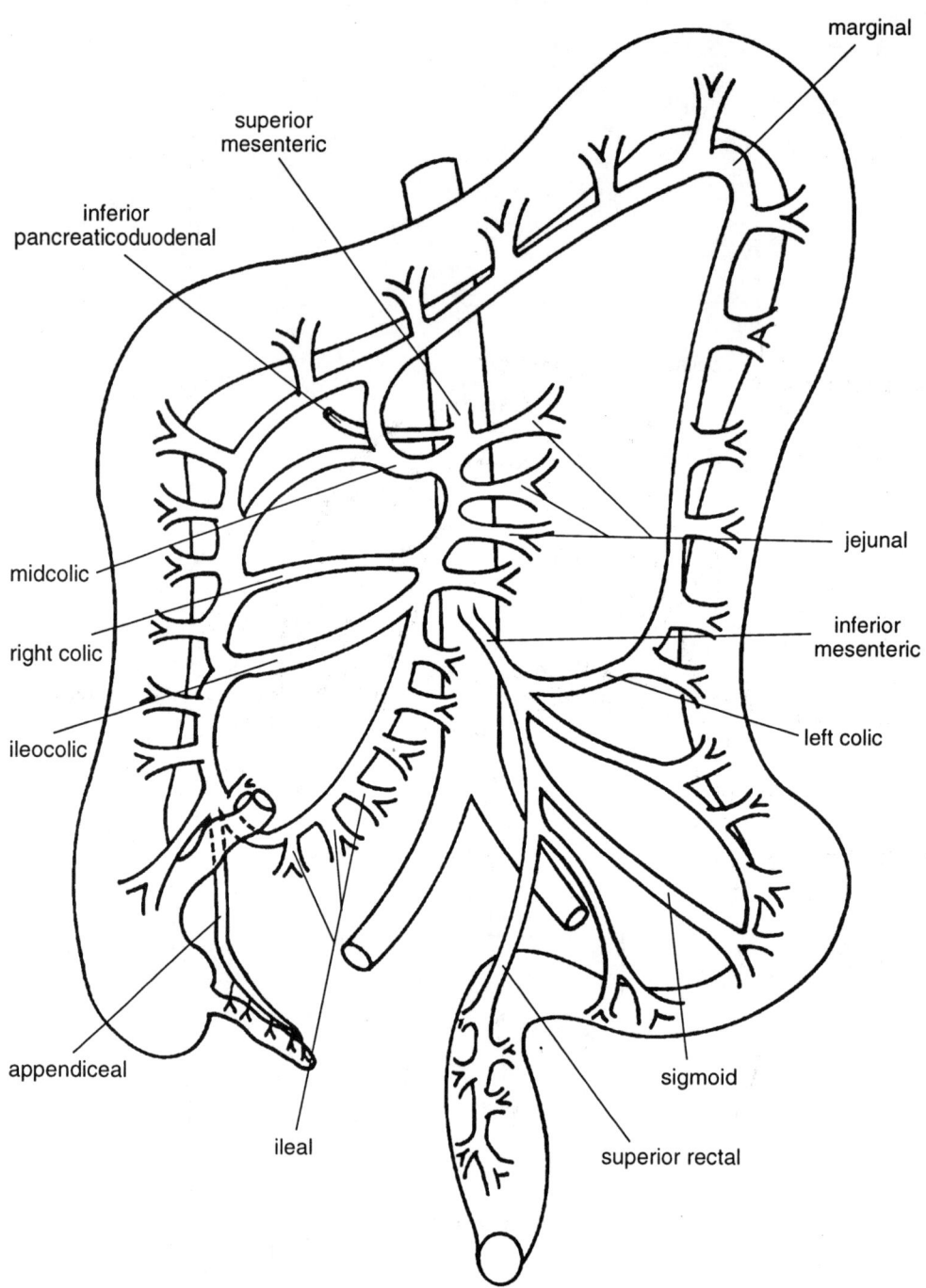

Superior and inferior mesenteric arteries (transverse colon raised)

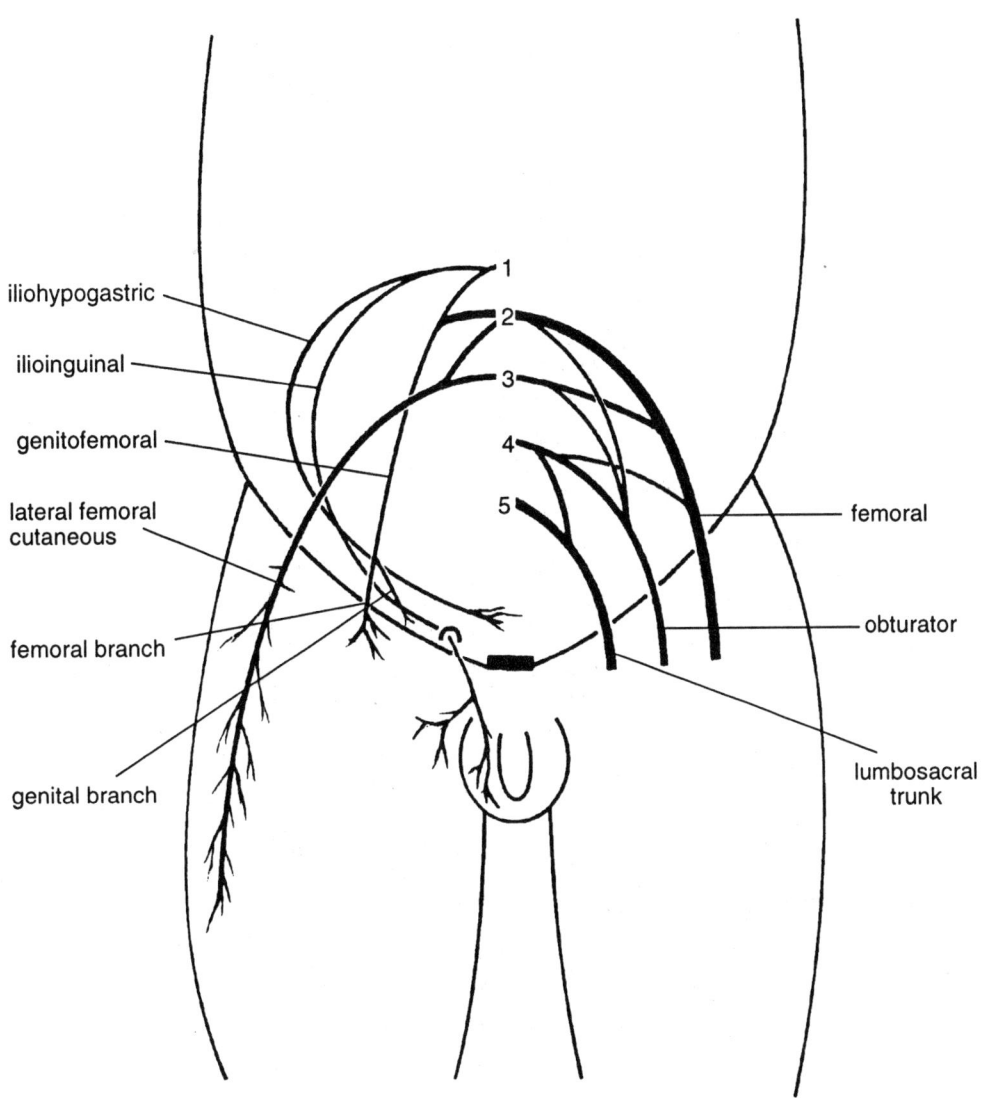

iliohypogastric

ilioinguinal

genitofemoral

lateral femoral
cutaneous

femoral branch

genital branch

1
2
3
4
5

femoral

obturator

lumbosacral
trunk

Lumbar plexus

Abdomen Self-Assessment

1. In the inguinal region, one statement is incorrect:

   A The deep layer of the superficial fascia is a relatively strong layer.
   B The transversalis fascia is attached to the inguinal ligament.
   C The transversalis fascia forms the floor of the inguinal canal.
   D The superficial layer of the subcutaneous fascia is unimportant as regards the strength of the abdominal wall.
   E The deep layer of the subcutaneous fascia is attached to the fascia lata of the thigh.

2. Regarding the inguinal canal, one statement is incorrect:

   A The falx inguinalis is formed by the merged free borders of the transversus abdominis and the internal oblique muscles and aponeuroses.
   B Inguinal herniation medial to the inferior epigastric artery may well contain a portion of the urinary bladder which lies directly posteriorly.
   C Inguinal hernias based on the embryological descent of a peritoneal processus vaginalis are invariably located within the confines of both internal and cremasteric fascial layers.
   D Direct inguinal hernias, but not indirect inguinal hernias, occur in women as they do not experience gonadal descent.
   E Normally, the processus vaginalis is represented in the adult male as the tunica vaginalis of the testis. All other portions disappear.

3. Regarding vestigial remnants in the fully developed adult, one statement is correct:

   A The medial unbilical ligament represents the remnant of the fetal umbilical artery.
   B The urachus is the remains of the fetal duct draining the colon.
   C Meckel's diverticulum is an outpouching of the hindgut.
   D The falciform ligament is all that is left of the pericardioperitoneal membrane.
   E The vermiform appendix is a remnant of the allantois.

4. Regarding the peritoneal cavity, one statement is incorrect:

   A The uterine tubes are the sole openings of the peritoneal cavity to the outside world.
   B The greater omentum constitutes the primitive dorsal mesentery of the transverse colon.
   C The hepatorenal and rectouterine pouches are the most dependent portions of the peritoneal cavity in the supine female. They are therefore frequent sites of fluid collection.
   D Peritoneal fusions develop between juxtaposed parietal and visceral peritoneum along ascending and descending colon, pancreas, duodenum and dorsal mesentery of the spleen. The fused areas may be opened without bleeding to mobilize any of these organs.
   E Peritoneal gutters alongside the ascending and descending colon connect subphrenic and pelvic cul-de-sacs, and direct fluid flow in the peritoneal cavity.

5. With obstruction of the portal vein, one statement is incorrect:

    A The spleen will uniformly be enlarged.

    B Portal venous blood will be returned to the right side of the heart by collateral channels such as the azygos system.

    C Hepatic arterial flow will increase as portal obstruction increases.

    D Esophageal hemorrage from the umbilical caput medusa is a common serious complication.

    E Esophageal varices are frequently seen.

6. Regarding the lesser sac, one statement is incorrect:

    A The site of attachment of the gastrohepatic omentum to the liver marks the site of the bare area between liver and diaphragm.

    B The anterior sac wall is largely stomach.

    C The pancreas lies in the sac floor (dorsal aspect).

    D May be entered through embryonic remnants of the anterior and posterior mesenteries.

    E The embryonic dorsal (posterior) mesentery forms part of the left lateral wall.

7. Regarding the autonomic nervous system of the gut, one statement is incorrect:

    A Splanchnic nerves consist of preganglionic sympathetic fibers from T5–L2.

    B The celiac ganglion offers postganglionic neurons for the parasympathetic preganglionics of the vagus nerves.

    C Thoracic splanchnic nerves enter the abdomen through the crura of the diaphragm.

    D The neurons of the sympathetic trunk play no active role in the innervation of the gut.

    E Both sympathetic and parasympathetic fibers reach the liver on the hepatic artery wall.

Match each numbered item with a letter on the diagram.

  8. Caudate lobe.

  9. Porta hepatis.

10. Ductus venosus.

11. Ligamentum teres.

12. Gallbladder.

13. Inferior vena cava.

14. Right lobe.

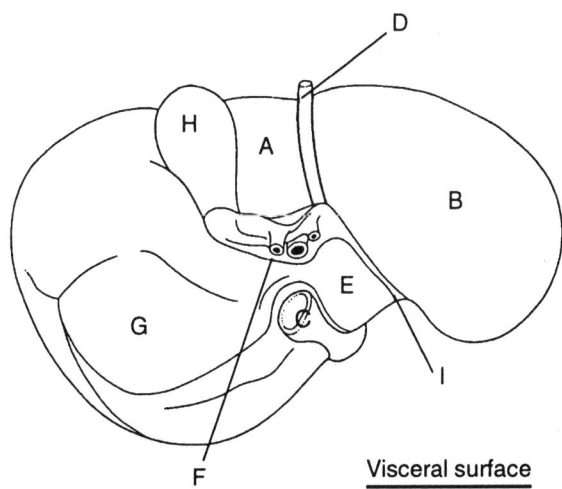

Visceral surface

15. Regarding the pancreas, one statement is correct:

    A Its tail is related to the hilum of the spleen.
    B Its body demonstrates a groove on its caudal surface for the splenic vessels.
    C The celiac axis arises posterior to its neck.
    D As an extraperitoneal structure, its blood supply is derived from the extraperitoneal renal arteries.
    E It is totally developed within the leaves of the dorsal mesentery.

16. Which of the following statements regarding the spleen is correct?

    A Lies in the midaxillary line.
    B Is not palpable unless pathologically enlarged.
    C Receives arterial blood from the superior mesenteric artery unless anomalous vessels are present.
    D Venous return is to the inferior vena cava.
    E Due to peritoneal fusions occurring during its development, it becomes a 'secondary' retroperitoneal structure in the adult.

17. Regarding the gallbladder, one statement is incorrect:

    A If the cystic duct becomes obstructed by a stone, jaundice will become apparent.
    B Arterial supply arises from the right or left or common hepatic artery.
    C Its bed forms the caudal portion of the right sagittal fissure.
    D May produce a stone that erodes into the hepatic flexure area of the colon or into the duodenum.
    E Its cystic duct usually drains into the right hepatic or common hepatic bile duct.

18. One of the following statements regarding the liver is incorrect.

    A Is formed between the leaves of the ventral mesentery.
    B Receives both portal vein and systemic arterial blood.
    C Contains hepatic veins which drain into the azygos system.
    D Is an intraperitoneal structure, although showing a broad area posteriorly devoid of peritoneum.
    E Its coronary ligament surrounds the bare area and represents the reflection of the ventral mesentery onto the parietal peritoneum.

19. Concerning the urinary tract, one of the following statements is incorrect:

    A Kidneys normally extend from T12–L3.
    B The ureter is totally extraperitoneal and is found crossing the psoas muscle from its lateral to its medial side.
    C The ureter constantly enters the pelvis over the bifurcation of the aorta.
    D Gonadal vessels cross ventral to the ureters.
    E Ureteral blood supply comes from both renal and vesical arteries.

20. Regarding the circulation of the intestinal tract, one of the following statements is incorrect:

    A   Circulation supplied by the inferior mesenteric artery is supported by the superior mesenteric artery through the marginal artery (Drummond) and *vice versa.*
    B   Vasa recta to the jejunum are long compared with those of the ileum.
    C   In the event of obstruction of the hepatic artery proximal to the gastroduodenal branch, liver necrosis usually occurs.
    D   The superior mesenteric artery arises posterior to the pancreas, but runs ventral to the third portion of the duodenum.
    E   The celiac axis provides arterial blood to the entire foregut.

21. Regarding the lumbar plexus, one of the following statements is incorrect:

    A   Innervates the psoas major muscle.
    B   Does not contain dorsal primary rami of its constituent nerves.
    C   The psoas muscle shelters under its medial border the obturator nerve and, under its lateral border, the femoral nerve.
    D   The ilioinguinal nerve arises from L1. It is at risk during inguinal hernia operations.
    E   The fifth lumbar nerve contributes to both the femoral and obturator nerves, and to the lumbosacral trunk.

22. Concerning the diaphragm, one of the following statements is incorrect:

    A   The diaphragm is formed entirely by the septum transversum.
    B   Congenital hernias through the diaphragm due to failure of closure of the pericardioperitoneal membranes pose serious respiratory problems for the newborn.
    C   There are three openings through the diaphragm: the aortic hiatus at T12; the esophageal hiatus at T10; and the opening for the inferior vena cava at T8.
    D   The thoracic duct passes through the aortic hiatus.
    E   The peritoneal surface of the diaphragm as well as the pleural surface has sensory innervation supplied by the phrenic nerve.

Answers and Explanations

1. **B** The transversalis fascia is attached to the ileopectineal ligament. It is readily separated from the inguinal ligament. It forms the floor of the inguinal canal - the first abdominal wall layer to be penetrated by the descending testis. The deep layer of subcutaneous fascia does indeed merge with the fascia lata of the thigh. This will be found to be an important anatomical fact when discussing extravasation of urine in the next section. The superficial layer of the subcutaneous fascia is little but fat.

2. **D** Females do produce a gubernaculum and a processus vaginalis in a cramped inguinal canal. Congenital and acquired indirect inguinal hernias are not rare in the female. Direct hernias, however, are most unusual.

3. **A** The urachus is a remnant of the allantois which originally drained the cloaca, and later the bladder. Meckel's diverticulum is found in the distal midgut. It is a remnant of the omphalomesenteric duct (to the yolk sac). The pericardioperitoneal membranes close in the posterior parts of the diaphragm and are not concerned with the primitive mesenteries. The appendix has no connection with the allantois.

4. **B** The greater omentum fuses with the 'regular' dorsal mesentery of the transverse colon, but contains no vessels or nerves supplying the transverse colon. It is readily separated from the transverse mesocolon. The celomic derivatives - pericardial, pleural, and peritoneal cavities – normally allow no communication with the outside world save via the fallopian tubes.

5. **D** When relatively low-pressure portal blood has trouble getting through the liver, the spleen will enlarge due to back pressure, and alternate routes such as the azygos system will be exploited. Meanwhile, the hepatic artery flow increases in a reflex effort to supply the liver cells. Esophageal varices and esophageal hemorrhage are common with portal obstruction. The formation of a caput medusa about the umbilicus is evidence of another alternate route for portal blood in the presence of portal vein obstruction. These enlarged collateral vessels, however, do not receive enough blood or pressure to blow up and burst as do the esophageal varices.

6. **A** The site of attachment of the gastrohepatic omentum (ventral mesentery) to the liver marks the course of the ductus venosus from liver to inferior vena cava. This site bears no relationship with the bare area of the liver. The lesser sac is readily entered through the epiploic foramen of Winslow or through the gastrohepatic omentum (ventral mesentery). Also, easy entrance is obtained through the gastrosplenic or gastrocolic parts of the dorsal mesentery. The splenorenal ligament (dorsal mesentery) forms (with the splenic hilum) the left lateral wall of the sac.

7. **B** Vagal preganglionics meet postganglionic neurons in the walls of the viscera. The sympathetic trunks perform no active role in the innervation of the viscera. They distribute postganglionic fibers to the segmental spinal nerves. Sympathetic fibers for the gut become postganglionic by synapse with neurons in preaortic ganglia.

Autonomic fibers by and large reach their visceral targets by traveling on appropriate arteries.

8. **E**

9. **F**

10. **I**

11. **D**

12. **H**

13. **C**

14. **G**

15. **A** The splenic vessel groove is on the cranial pancreatic border. The celiac axis arises cranial to the pancreas. The pancreas is 'secondarily' extraperitoneal due to fusion of one leaf of its mesentery with parietal peritoneum. Its major blood supply remains the artery of the foregut – the celiac axis. The pancreas develops by buds into both dorsal and ventral mesenteries, the ventral part swinging around behind the duodenum to join the dorsal part.

16. **B** The spleen lies deep and far posteriorly in the scapular line (medial border of scapula). It is not palpable in this normal site, unless pathologically enlarged. It is developed in the foregut dorsal mesentery, and is nourished by the foregut celiac axis. Venous return enters the portal system. It remains intraperitoneal though fusion 'eats up' a good deal of its mesentery.

17. **A** Obstructive jaundice will only occur with obstructive difficulties of the bile ducts draining the liver proper. The cystic artery and duct are variable in their anatomy - almost anything may be found. By virtue of its close relation to both duodenum and hepatic flexure, erosion and passage of a large stone into either of these structures is not unknown.

18. **C** Hepatic veins drain into the inferior vena cava. The azygos is connected with the portal not the hepatic veins.

19. **C** The landmark for ureteral location is the bifurcation of the common iliac artery. The ureter passes right over it.

20. **C** The gastroduodenal artery is connected with the superior mesenteric system via the pancreaticoduodenal arcade. Collateral blood supplied by this route will keep the liver cells happy.

21. **E** L5 runs entirely to the sacral plexus via the lumbosacral trunk. Dorsal primary rami innervate back muscles and skin. The lumbar plexus is composed entirely of ventral primary rami.

22. **A** Embryonic pericardioperitoneal membranes, the esophageal mesentery, and the body wall all assist the septum transversum in forming parts of the diaphragm. Large congenital diaphragmatic hernias permitting entry of abdominal structures into the pleural cavity may prevent development of the lung and/or seriously interfere with

respiration in the newborn. As the peritoneal surface of the diaphragm is phrenic nerve innervated, irritation of diaphragmatic peritoneum, as with irritation of diaphragmatic pleura, may be demonstrated as pain in the shoulder top. The phrenic nerve originates largely in C4 nerve, and refers pain stimuli from the diaphragm to the C4 dermatome.

 # Pelvis and Perineum

Bones and ligaments. The pelvis consists of the hip bone (fused ilium, ischium and pubis), and the posterior sacrum and coccyx. Be aware of the structure of the hip bone, the terms applied to the various parts of the three component bones and the way they come together to form the acetabulum, and the various characteristic prominences that make it possible to describe accurately any point on the bony pelvis. This is essential for accurate communication. Using a specimen of the bony pelvis, identify the following:

Bony boundaries of the true and false pelvis, linea terminalis, and arcuate and iliopectineal lines;

Anterior superior and inferior, and posterior superior and inferior, iliac spines;

Body, ramus, tuberosity and spine of the ischium, and the greater and lesser sciatic notches;

Body, inferior and superior rami of the pubis, pubic tubercle, pecten pubis (pectinate line), iliopubic eminence and obturator groove;

Obturator foramen and obturator groove. What neurovascular bundle passes through the groove?

Visualize the sacrotuberous and sacrospinous ligaments, and the resultant greater and lesser sciatic foramina. What structures run through each of these foramina? The answer: everything passes through the greater sciatic foramen except obturator internus ('everything' includes piriformis, superior and inferior gluteal vessels and nerves, internal pudendal vessels and nerve, sciatic and posterior femoral cutaneous nerves, and nerves to quadratus femoris and obturator internus). The obturator internus nerve and the pudendals pass out of the pelvis through the greater foramen, and reenter through the lesser sciatic foramen.

The obturator internus tendon alone uses only the lesser sciatic foramen.

Visualize the obstetrically important diameters of the pelvic inlet (conjugate vera) and outlet. The pelvic inlet is measured from the sacral promontory to the mid-superior pubic symphysis. The pelvic outlet is measured from the interior symphysis to the coccyx.

Peritoneum. Confirm for yourself that the rectovesical pouch (in men) and the rectouterine pouch (in women) are located well posterior and cannot be palpated by examination through the abdominal wall, and that either structure is easily reached via digital examination of the rectum and/or vagina. Understand the route taken by fluids from the abdomen to these pouches. Which other deep peritoneal pouch, unapproachable by abdominal physical examination, collects intra-abdominal fluid? [The answer: hepatorenal (Morison's) recess.] Understand that the peritoneum reflects onto the proximal rectum (rectosigmoid) from the rectovesical or rectouterine pouch. Thus, the rectum is extraperitoneal caudal to the level of the pouch.

85

Rectum.   The major blood supply of the rectum is the superior rectal artery, the terminal branch of the inferior mesenteric artery. The embryological urorectal septum divides the cloaca into rectal and urogenital halves. There are no blood vessels or nerves running across the septum between these two halves. Similarly, there are no vessels or nerves running to the rectum from the posterior-lying sacrum. Thus, the rectum is readily separated from urogenital structures anteriorly and presacral fascia posteriorly. Freed in this fashion, the rectal blood supply from above is elevated with the rectum. Use of these fusion planes renders resection of the rectum a safe and virtually bloodless procedure.

The so-called lateral ligaments of the rectum are thickenings of fibrous tissue along the lateral borders of the retroperitoneal rectum which carry autonomic nerves to the rectum, and the small middle rectal artery and vein to and from the internal iliac. Lymphatics, too, may take this route from rectum to iliac nodes, although the more common route is through the lymphatics in company with the superior rectal vessels.

Urinary tract in the pelvis.   Review the description of the ureter and its abdominal course and blood supply in the Abdomen section (see page 63). The deep-lying ureter is dorsal to the gonadal and uterine vessels. After entering the pelvis over the bifurcating common iliac arteries, the ureter pursues its extraperitoneal course around the pelvic bowl to reach the bladder. In the male, it is crossed, just before it enters the bladder, by the vas deferens as the latter progresses to the prostate. The ureter pursues a tangential 2-cm course through the bladder wall and finally enters the bladder at the lateral angle of the bladder trigone. In the female, the pelvic ureter is crossed anteriorly in its distal portion by the uterine artery, then runs forwards alongside the supravaginal cervix to the bladder.

Bladder.   This is held in place by stout pubovesical ligaments (or puboprostatic ligaments in men). The retropubic space between bone and bladder (space of Retzius) is alive with veins, as urologists well know! Peritoneum covers the bladder superiorly whereas inferiorly the bladder is extraperitoneal. From the bladder apex at the pubic symphysis, the median umbilical fold of peritoneum (over the urachus) runs to the umbilicus. The medial peritoneal umbilical folds rise from the bladder peritoneum to the umbilicus lateral to the median fold. They apparently represent the obliterated umbilical arteries, the continuation of the hypogastric (internal iliac) arteries beyond the take-off of the superior vesical arteries.

As the bladder fills with urine, it ascends the abdominal wall in the extraperitoneal fatty plane. It often appears in the medial wall of direct inguinal hernias as it comes to lie under Hesselbach's direct inguinal triangle. The full bladder is palpable in the suprapubic abdomen. Blood supplying the bladder is furnished by vesical branches of the hypogastric artery; the nerves to the bladder are primarily parasympathetics from S3-4 and the pelvic plexus as well as sympathetics extending downwards to the pelvic plexus from the hypogastric plexus over the bifurcating aorta. The orifice of the bladder is the urethra. In the male, the urethra immediately enters the prostate; in the female, the urethra enters the urogenital diaphragm. The urethral orifice in both sexes forms the apex of the bladder base trigone. The ureteral orifices form the two base points of the triangle.

Male internal genitalia.

Prostate. This is the walnut-sized firm, smooth, musculoglandular structure lying between the bladder apex and urogenital diaphragm in men. It rests posteriorly on the rectum (through which it is readily palpated on digital rectal examination), and is pierced by the urethra and the ejaculatory ducts. The prostate is held in place by the puboprostatic ligaments, the urogenital diaphragm and the embracing arms of the puboprostatic portion of levator ani. While recalling the passageway through the cloaca of the urorectal septum separating hindgut from anterior urogenital stuctures (bladder, seminal vesicles, prostate), be aware that the short distance between the prostate and rectum is not traversed by vessels or nerves. The rectum and prostate may thus be surgically separated bloodlessly by blunt reopening of the urorectal septum.

The prostatic glands produce a secretion which is part of the semen ejaculate. It is presumably involved in the health, welfare and function of the sperm it contains.

Seminal vesicles. These paired coiled gland-lined tubes of around 5 cm in length are found in the male attached to the posterior aspect of the bladder resting on the rectum and separated from the latter by remnants of the urorectal septum. At their deepest extent, the vesicles abut the base of the prostate. The vasa deferentia pass to the prostate between the two vesicles and, at the prostate, the efferent ducts of the vesicles join the vasa to form the ejaculatory ducts, which traverse the prostate to enter the prostatic urethra. The seminal vesicles contribute a large part of the volume of the ejaculate, supplying nutrition, buffering and other materials necessary for sperm health and function. (Sperm are stored not in the vesicles, but in the epididymus and vas.)

Prostatic urethra. The male urethra has three parts: prostatic, membranous, and cavernous (penile). (The latter two are described with the Perineum.) On opening the prostatic urethra, a longitudinal ridge (urethral crest) is seen elevating the posterior urethral wall. In the midportion of the ridge is a rounded swelling (seminal colliculus or verumontanum). The two ejaculatory ducts open into the urethra on either side of the collicular summit. At the summit is a single opening (prostatic utricle or uterus masculinus), the male remnant of the united ends of the müllerian ducts, which form the uterus and vagina in the female. Numerous small orifices of prostatic gland ducts are seen in the floor of the urethra on either side of the urethral crest.

(The female urethra, sans prostate, runs directly from the bladder into the urogenital diaphragm, imbedded in the anterior wall of the vagina.)

Female internal genitalia.

Ovaries. Homologues of the testes, these are suspended by a peritoneal mesenteric fold from the posterior aspect of the broad ligament, the mesovarium. The lateral ovarian extremity is suspended by a fibroperitoneal fold encasing the ovarian vessels, the suspensory ligament of the ovary (or infundibulopelvic ligament). Remember that the ovarian arteries stem from the aorta, and that the ovarian veins run to the inferior vena cava on the right and to the renal vein on the left. Extending from the medial ovarian aspect is the homologue of the male gubernaculum, the ligament of the ovary. This structure runs in the broad ligament to the uterus, then caudally in the broad ligament as the round ligament of the uterus. Reaching the pelvic well, it ascends to the deep inguinal ring of the transversalis fascia. Traversing the ring, it passes through the inguinal canal to end in the tissues of the labium majus. As in the male embryo, a peritoneal processus vaginalis accompanies the round ligament (gubernaculum of the male) into the inguinal canal. If this processus persists in the female, it is termed the canal of Nuck (Dutch anatomist; 1650–1693) and may lead to indirect inguinal hernias in women as it does in men.

Uterine (fallopian) tubes. These extend from the superior lateral angles of the uterus laterally to the ovaries. The tubes open into the cavity of the uterus and convey ova from the ovaries to the uterine cavity. The open ends of the tubes are expanded and terminate in finger-like fimbriae which 'clasp' the ovaries. The tubes are covered by the peritoneum of the cranial termination of the broad ligament.

The fallopian tubes, as they open into the peritoneal cavity and uterus-vagina, constitute the only opening from outside to inside of the body cavity.

Uterus. Thick-walled, muscular and pear-shaped, this organ resides between bladder and rectum, with the uterovesical peritoneal pouch anteriorly and the deep, clinically important, rectouterine pouch (of Douglas) posteriorly. The uterus consists of a body and a cervix. The uterine body cranial to the entrance of the tubes is termed the uterine fundus. The body is peritoneum-covered. Laterally, extending from the anterior and posterior surfaces of the uterus, is a double fold of peritoneum, the broad ligament, which merges with the parietal peritoneum at the pelvic walls. Between the peritoneal layers of the broad ligament lie the fallopian tubes, ovarian ligaments and round ligaments and, at the uterine border, the vessels and nerves of the uterus.

The cervix is the thick terminal portion of the uterus. The supravaginal part is extraperitoneal; the vaginal part extends into the vagina where the cavity of the uterus ends as the external os (ostium). (The internal os is the entry of the uterine cavity into the narrower cervical canal.) The external os is surrounded by the swollen 'lips' of the cervix.

Uterine ligaments. These are important structures capable of holding the uterus in position despite the mechanical rigors of pregnancy. The most important ligaments are the transverse cervical (cardinal) ligaments and the uterosacral ligaments, stout condensations of subperitoneal connective tissue. The transverse cervical ligaments run in the base of the broad ligament from cervix to lateral pelvic walls – the fascia over the levator ani muscles. The uterosacral ligaments pursue a curved course along the pelvic brim from the cardinal ligament to the periosteum of the sacrum. Other important supporting structures are the pelvic diaphragm (levators ani) and the urogenital diaphragm. The broad and round ligaments are not important supporting structures.

Vessels and nerves of the uterus. The uterine artery is a large branch of the internal iliac which reaches the uterus via the transverse cervical ligament. Just before reaching the uterus, it passes across (ventral to, in front of) the ureter as the latter angles up towards the bladder base. A cardinal surgical sin is ligation of the ureter in company with the uterine artery in the common operative procedure of uterus excision (hysterectomy). The uterine artery and the ureter lie **close** to each other at the site of necessary ligation of the artery. The uterine artery sends vaginal artery branches downwards before it ascends along the sides of the uterus between the layers of the broad ligament.

Autonomic nerves reach the uterus from the pelvic plexus lying in the retroperitoneum of the pelvic walls. The nerves follow the same lateral-to-medial path along the transverse cervical ligament as does the uterine artery.

Nerves in the pelvis.

Lumbar plexus (from above the pelvis):

Femoral nerve: Along the **posterolateral** border of psoas, across 'false' pelvis, under inguinal ligament to the femoral triangle;

Obturator nerve: Along **posteromedial** border of psoas, through obturator foramen (in the obturator canal with the obturator vessels) to thigh under pectineus muscle.

Sacral plexus:

Lumbosacral trunk, L4–5, deep to obturator nerve, joins sacral plexus. L4 contributes to both sacral and lumbar plexuses [= 'furcal (forked) nerve'] whereas L5 contributes only to the sacral plexus;

S1–4 with the lumbosacral trunk (L4–5) form the sacral plexus whence the **sciatic** nerve issues. The sciatic nerve has all root levels, exits the pelvis from beneath the lower border of piriformis, through the greater sciatic foramen, to the posterior midline of thigh. It is accompanied by the S1–3 posterior thigh skin nerve, the posterior femoral cutaneous nerve.

In the pelvis, the sacral plexus lies against the pelvic wall over piriformis, and innervates **all** muscles running from pelvis to hip, the muscles of the levator bowl, the skin over the caudal buttock and skin down the back of thigh (posterior femoral cutaneous). It forms the pudendal nerve, which supplies skin and muscles of the perineum (including the external anal sphincter), and the genitalia. (The pudendal nerve is described in the Perineum section.)

Autonomic nerves:

Sympathetic trunk and sympathetics to peripheral nerves: The lumbar sympathetic trunk continues down over the surface of the sacrum running medial to the anterior sacral foramina which transmit the sacral nerve roots. The sacral trunk contains fibers from lumbar cord segments L1–2 which entered the lumbar trunk via white rami communicantes in that lumbar region. In the sacrum, these fibers synapse in the sacral ganglia of the trunk, then enter sacral roots and the sciatic nerve via gray rami communicantes as postganglionic sympathetic supply to smooth muscle (mostly blood vessels) and glands of the lower extremity;

Sympathetics to viscera: These use midline plexuses and ganglia. Lumbar splanchnics to and through the sympathetic trunk without synapse progress to the inferior mesenteric plexus (*cf* thoracic splanchnics forming more cranial plexuses). Fibers bound for the pelvis continue through the inferior mesenteric plexus to the hypogastric plexus lying over the sacral promontory between the common iliacs. From there, they progress down the walls of the pelvis (hypogastric nerves) to join the pelvic plexus lying over the coccygeus muscle in the true pelvic wall. (The sciatic nerve is not in the way as it leaves the pelvis over the superior border of coccygeus). Postganglionics from here run to the pelvic viscera and genitalia along visceral branches of the internal iliac and internal pudendals;

Parasympathetics (sacral part of craniosacral outflow): Parasympathetic fibers from sacral nerve roots 2–4 branch off and run to the pelvic plexuses as 'pelvic nerves', 'pelvic splanchnics' or 'nervi erigentes'; all describe the same fibers. They run through the plexus, accompany postganglionic sympathetics to viscera via visceral arteries, and finally synapse in ganglia in the walls of the viscera and genitalia they supply. Visualize these autonomics, sympathetic and parasympathetic, reaching the rectum, bladder and uterus from the lateral pelvic wall in the lateral supporting ligaments (lateral ligaments of the rectum, transverse ligament of the uterus, pubovesical ligaments of the bladder). This is the same avenue used by the blood vessels.

Bladder and rectal function. These functions in the adult are complex. At the time of birth, certain reflexes are already established. The rectum responds with a defecation reflex to the pressure of stool delivered by the sigmoid colon. Similarly, the bladder responds to the filled tension of urine by a micturition reflex. As we grow older, we learn to control these important reflexes through participation of the brain itself. A well-functioning human adult is continent, voiding and defecating at appropriate times and places. The anatomy of these reflexes involves both autonomic systems and smooth muscle, plus the central nervous system and striated muscle. Orchestration requires intact sensory nerves and spinal cord with integrating connections to and from the brain stem and cerebrum.

Micturition. Bladder distention is signaled via afferents retracing both sympathetic and parasympathetic routes back to the spinal cord. The reflex center lies in neurons supplying the sacral parasympathetics – S2–4, the pudendal and levator ani nerves – also S2–4, and sympathetics from L1–2. (Knowing that the cord ends at the junction of the L1–2 **vertebrae**, it is apparent that to reach the L1–2 **cord level** requires a rise to the T11–12 vertebral level. Thus, the reflex arms to and from both bladder and rectum cover a distance of around 25–30 cm.) Within the cord, the distention reflexively stimulates both parasympathetic and sympathetic preganglionic neurons. Parasympathetic efferents run to the bladder via the cauda equina, pelvic plexuses and lateral ligaments of the bladder. Sympathetic efferents from L1–2 travel via lumbar splanchnics to preaortic ganglia, hypogastric plexuses, pelvic plexuses and lateral ligaments of the bladder. In addition, the pudendal nerve perineal branch (S2–4) to the sphincter urethrae of the urogenital diaphragm and the nerves to levator ani (S4) must all recognize the distention message. When everything is in place, the bladder neck relaxes (sympathetic), the bladder detrusor muscle contracts (parasympathetic), the levator and urethral sphincter relax (pudendal and levator nerve) and urine is expelled. Conscious control of urination resides in the cerebrum where the micturition distention signal is relayed and perceived, and conscious contraction of the urethral sphincter is available to interrupt the reflex until a proper locale is reached.

Defecation. This is similar to micturition. Distention of the rectum is reported to the cord. Sympathetic and parasympathetic stimuli are returned to the pelvic plexus and out to the rectum via its lateral ligaments. In addition, autonomics reach the internal anal smooth muscle sphincter, stimuli reach the external anal sphincter in the pudendal nerve and levator ani via its nerve from the sacral plexus and, through a similar contraction-relaxation action, stool is expelled. Conscious control is learned and effected through external sphincter and levator ani resistance to the contracting rectal pressure.

As long as the reflex arc is intact, micturition and defecation will take place in response to bladder or rectal distention; control is lost with spinal cord injury above the reflex level – for example, above the T11 **vertebral** level. In this situation, automatic emptying of the bladder and rectum occur in the same way an untrained infant functions. Injury below the reflex level interrupts the efferent and/or afferent arms of the reflex loop, and destroys even the reflex performance. 'Overflow' dribbling of urine and feces results. Such lesions include low spinal cord injury and lesions of the cauda equina.

Genital function.

Again, the nerve supply requires afferents and efferents, both visceral and somatic, to and from the genital organs as with the bladder and rectum. Afferents to the lower cord, with connections to the brain, and efferents both somatic (pudendal) and sympathetic and parasympathetic to the pelvic plexus are employed.

Male. Genital outflow from the pelvic plexus passes along the lateral ligaments of the rectum and bladder, sending branches to the vas, vesicles and prostate. The outflow continues alongside the prostate and out under the pubic arch to the cavernous tissue of the penis. Sympathetic stimulation of the smooth muscle of the vas introduces sperm into the prostatic urethra accompanied by sympathetic stimulated secretions from the vesicles and prostatic glands. This constitutes the few milliliters of ejaculate. Cavernous bodies become filled and distended with blood (erection) by a combination of parasympathetic-directed penile arterial dilation and venous occlusion. Stimulation for ejaculation occurs as the semen reaches the penile urethra. At this point, through pudendal nerve activity, spasmodic contractions of striated bulbospongiosus muscle expels the semen. Again, visceral and somatic afferents and efferents are required. Cerebral activity participates in, and controls, the performance to a degree.

Female. The totally similar anatomy produces clitoridal erection, vaginal secretion and vaginal contractions during sexual intercourse. Again, difficulties anywhere along the line may cause problems. Erection is actually possible in rare instances with injuries above T10 if the autonomic reflexes remain functionally intact.

Muscles of the pelvic bowl.

The piriformis, coccygeus and obturator internus muscles line the lateral walls of the pelvic bowl. Piriformis originates from the sacrum and passes from the pelvis through the greater sciatic foramen to insert on the greater trochanter of the femur. Obturator internus arises from the rami of pubis and ischium, and from the obturator membrane, the fascia covering the bony obturator foramen, and bends at a right angle laterally under the ischial spine to leave the pelvis through the lesser sciatic foramen to finally reach and insert on the greater trochanter. Both muscles rotate the thigh laterally. Piriformis also abducts the thigh. Slung from the lowermost part of the bowl, the levators ani from each side stretch across the open-ended bowl to meet in the midline and thus close off the bowl. This diaphragm originates from the posterior aspect of the lower pubic symphysis, then across obturator internus (originating from the fascia covering the surface of this muscle) to the ischial spine. There is a gap in this hammock between the pubic origins of the muscle. Through this gap run the urethra, vagina and anorectum. Thus, the pelvic diaphragm closes off the abdominopelvic cavity below as the respiratory diaphragm closes off the cavity above. The levator supports the viscera passing through it as the two sides of the pubococcygeus meet in the midline behind the prostate (or vagina) and the rectum to form U-shaped slings for these structures, the levator prostatae, pubovaginalis and puborectalis muscles. The U-shaped part of puborectalis merges with the external anal sphincter laterally and posteriorly, and is an important part of the rectal sphincter continence apparatus. The coccygeus portion of the levator ani is a vestigial remnant of an important muscle helping to operate the tail in lower animals. In humans, it consists of a few strands of muscle resting on the sacrospinous ligament.

Blood vessels of the pelvis.

The aorta bifurcates at level L4, which is well above the pelvis proper.

Common iliac arteries: These run over the common iliac veins as the latter lead up to the vena cava lying alongside the aorta. Thus, these iliac veins are always posterior and slightly medial to the corresponding arteries ('Prince Philip and the Queen'). Common iliac arteries bifurcate at the brim of the true pelvis to become the external and internal iliacs. Invariably, immediately in front of this bifurcation lies the ureter. (The internal iliac artery is often called the hypogastric artery.)

External iliac artery and vein: These course on the surface of the psoas muscle, crossed by the gonadal vessels and the vas deferens or round ligament as they pass under the inguinal ligament into the femoral triangle. The artery lies lateral to the vein under the ligament. There are two important arterial branches of the external iliac in the abdomen: the inferior epigastrics and the deep circumflex iliacs. The inferior epigastrics form the lateral umbilical ligaments as they ascend to nourish the rectus abdominis muscle. The deep circumflex iliacs proceed laterally along the horizontal rami of the pubes to reach the anterior superior iliac spines and crests of the ilia. They anastamose with ascending branches of the lateral femoral circumflex and inferior gluteal arteries, and may prove to be an important source of collateral blood flow to the lower extremity in cases of femoral artery occlusion.

Internal iliac: Organize this according to its main branches:

Posterior parietal ($n = 4$): superior and inferior gluteal, iliolumbar and lateral sacral;

Intrapelvic visceral ($n = 3$; 4 with uterine): middle rectal, superior vesical (umbilical), middle and inferior vesicals, and uterine;

Caudal (distal) parietal ($n = 2$): obturator and internal pudendal.

Memorize the positions of these:

Superior gluteal: Exits posteriorly to reach the buttock over the cranial border of piriformis, between this muscle and gluteus medius;

Inferior gluteal: Exits posteriorly to buttock under caudal border of piriformis, between this muscle and coccygeus.

All of the above exit via the greater sciatic foramen.

Iliolumbar: The highest branch of the hypogastric artery, this primarily supplies the iliopsoas muscle and pelvic bone.

Lateral sacral: Enters sacral foramina, and supplies the contents of the sacral canal (cauda equina) and muscles of the dorsal sacrum.

Middle rectal vessels: These are small and unimportant. They reach the rectum via the lateral ligaments of the rectum.

Vesicle arteries: These spread out over the bladder. After giving off the superior vesicle branch, the internal iliac continues up the abdominal wall as a fibrous cord to the umbilicus, the obliterated umbilical artery (the medial umbilical ligament).

Uterine artery: Described above with the uterus.

Obturator: Runs straight down along the lateral wall of the true pelvis to exit through the obturator canal (filling the obturator groove of the pubic bone along with the accompanying vein and nerve) to the thigh beneath the pectineus muscle. Frequently, an accessory or replacing obturator artery arises from the inferior epigastric or external iliac artery and arcs downwards to pass through the obturator canal. Such an anomalous vessel may 'get in the way' during femoral hernia repair!

Internal pudendal: Exits the greater sciatic foramen with the inferior gluteals, then passes around the ischial spine and back into pelvis through the lesser sciatic foramen, running between obturator internus and levator ani. It then proceeds through a fascial canal across the surface of obturator internus to the ramus of the ischium. From there it ascends to the inferior ramus of the pubis and the symphysis in the area of the suspensory ligament of the penis or clitoris. Apparently a circuitous route, this is actually a straight line from its point of origin to the perineum. In its course across obturator internus, it is caudal to the take-off of levator from the obturator fascia, and thus runs in the lateral wall of the ischiorectal fossa and is technically already in the perineum as soon as it reaches the ischial spine. The pudendal nerve accompanies the internal pudendal artery and vein in this course (more on the pudendals in the Perineum section).

Perineum. This contains the outlets of the viscera of the abdominopelvic cavity: the urinary tract (urethra), genital tract (vagina in female, urethra in male), and intestinal tract (anal canal). The perineum consists of these tubes and the muscles that support and sphincter them, plus the external genitalia (labia and clitoris; scrotum, testes and penis). Thus, everything outside of the pelvic levator diaphragm is perineum. These outlets become outlets when they pass through the levator diaphragm. The levator is the inferior boundary of the abdominopelvic cavity; conversely, it is the upper cranial boundary of the perineum. The caudal boundary is the skin surrounding these outlets and extending to the pubic symphysis, arch of the ischiopubic rami, ischial tuberosities and coccyx. The sacrotuberous ligament overlaid by the margin of gluteus maximus defines the posterior boundary.

During embryological development, the cloaca is divided into anterior and posterior portions by the downgrowth of a fibrous urorectal septum. This septum divides and separates anterior urogenital parts from posterior anorectal parts. The septum reaches the perineal skin where it persists as the fibrous perineal body or midpoint of the perineum. The rectum and vagina use this structure for their anchoring support, and the perineal muscles and sphincters depend on this as an anchoring base. In the female, the perineal body alone is often referred to as the perineum by obstetricians and gynecologists, an indication of its importance in their business! Drawing a line across the diamond-shaped expanse of perineal skin from tuberosity to tuberosity and passing through the perineal body divides the 'diamond' into two triangles: an anterior urogenital triangle; and a posterior anal triangle.

These two triangles share blood and nerve supplies primarily from the internal pudendal vessels and the pudendal nerve. In addition, perineal branches of the posterior femoral cutaneous nerve from the sacral plexus, and the ilioinguinal nerve from the lumbar plexus, help supply the perineal skin.

Accurate anatomical knowledge of the perineum is absolutely essential for any clinician as problems in this area are legion. Accurate diagnosis and successful treatment are dependent on an expertise largely based on the anatomy of the area. Otherwise, clinical horrors such as urinary or rectal incontinence, chronic draining infections or obstetrical disasters may ensue.

Anal triangle. Contains the anal canal and its sphincters, and the fat-filled cavity through which the canal runs – the ischiorectal fossa.

Anal canal. The anterior part of levator ani runs from the posterior aspect of the pubic symphysis and, instead of continuing on to the coccyx, loops around the rectum (and prostate or vagina) as it enters the perineum before returning to the symphysis. This loop of levator – the puborectalis muscle – forms the proximal part of the anal sphincter mechanism. From here on out to the skin, the outlet of the intestinal tube is called the anal canal. The muscle loop is easily palpated posteriorly on rectal examination (anteriorly the loop is 'open' and cannot be felt). It merges with the intrinsic external anal sphincter. In its usual state of tonic contraction, the sphincter pulls the rectum forward against the symphysis, thereby closing the rectal lumen. This muscle loop is an important part of the fecal continence mechanism. (Similar loops around the prostate or vagina have similar, albeit less important, functions.)

Distal to and merging with puborectalis are the **external anal sphincter muscles**, which completely encircle the anus. They are attached anteriorly to the perineal body and posteriorly to a band of fibrous tissue running to the coccyx – the **anococcygeal raphe**. Anatomically in three parts, consider these and puborectalis as working together as voluntary muscles to keep the anal canal closed. This is accomplished both by the normal intrinsic tone of these muscles and by conscious voluntary contraction when necessary to resist the urge to defecate. When defecation is appropriate, the continent person can voluntarily relax these sphincters, allowing the defecation reflex to take over.

The **internal anal sphincter** is simply a thickening of the circular smooth muscle of the anal canal wall. It is overlaid by external sphincter striated voluntary muscle. The internal sphincter is involuntary. Its usual state is one of tonic contraction, which aids in keeping the anal canal closed but, with the onset of signals from the rectum above that it is time to defecate, the internal sphincter relaxes. It is at this stage that the voluntary sphincter muscles are consciously enlisted to take control of the defecation reflex in continent humans (animals).

The lining of the anal canal demonstrates the junction of the endoderm-lined posterior compartment of the cloaca (the end of the hindgut) and the ectoderm-lined (squamous cell) proctodeum. Marking the point where the two meet is the **pectinate** (anocutaneous or dentate) **line** ('toothed' or 'comb-like'). The vital importance of this line is that it marks the watershed between the portal-drained, superior rectal artery-nourished, insensitive (visceral afferents) rectum with its lymphatic drainage upwards along the rectal vessels, and the pudendal-drained and -nourished, exquisitely sensitive (somatic pudendal nerve) skin (anoderm) with its lymphatic drainage to the groin nodes. **Anal columns**, longitudinal folds of mucosa covering the terminal portions of the portal venous system – here enlarged and plexiform – extend to the pectinate line. Between the columns are thin mucosal folds termed **anal valves**, which subtend small recesses between the bulging columns called **anal crypts** or **anal sinuses**. Where the valves join the ends of the columns, the mucosa is raised into small **anal papillae**. The pectinate line is thus formed by the free margins of the valves and anal papillae. Tiny **anal glands** of uncertain function empty into the anal crypts. Accurate knowledge of this anal anatomy is vital to understanding the great number of anal complaints (for example, hemorrhoids, fistulae, abscesses, hypertrophied papillae and tumors) that you are likely to encounter daily as clinicians.

Ischiorectal fossae. These are fat-filled cavities on each side of the midline anal canal. Their boundaries are important as they are frequently the sites of large abscesses arising from the anal canal. They may harbor anal fistulae and are often involved in local trauma. The fossa roof on each side is the cranial boundary of the perineum, comprising the slanting levators as they descend to the sides of the anal canal and to the anococcygeal raphe from their take-off from the fascia overlying the midportion of the obturator internus muscle. The sides of the fossae are formed by the obturators internus caudal to the take-off of the levators. Medially is the anal canal, posteriorly are the glutei maximi with their free margins and inferiorly is the thick perineal skin. Inferior rectal vessels and nerves traversing the fossae are branches from the pudendals given off just anterior to the ischial tuberosities. After emitting these branches, the pudendals turn upward to run along the ischiopubic arches to the area of the urogenital triangle. Before reaching the tuberosities, the pudendals can be found running along the lateral wall of the fossae beneath a fold of fascia on the medial surface of the obturator internus muscles – the so-called Alcock's canal.

Urogenital triangle. This is the area between the ischiopubic arches and an imaginary line running between the ischial tuberosities and passing through the midpoint of the perineum (perineal body). There are two levels to this triangle: deep and superficial.

The deep part consists of a band or diaphragm of striated muscle stretching between the two arches, and covered above and below by layers of fascia – the superior (deep; inner) and inferior (superficial; outer) fascias of the urogenital diaphragm. The inferior layer is thicker and tougher, and is also termed the 'perineal membrane'. Recalling the descent of the urorectal septum to the perineum to divide the cloaca into anterior (urogenital) and posterior (anal) parts, be aware that through this anterior division, and thus through the urogenital diaphragm, must pass the outlets of the urinary and genital systems – the urethra in males, the urethra and vagina in females. The superficial part of the urogenital triangle consists of everything superficial to the inferior layer of the urogenital fascia: the external genitalia and terminal portions of the urinary tract. Its superficial boundary is the deep layer of subcutaneous fascia, called Colles' fascia here in the perineum. It is a continuation of Scarpa's fascia of the abdominal wall skin.

The deep part of the urogenital triangle is called the 'deep perineal pouch' and the superficial part is the 'superficial perineal pouch'.

Deep perineal pouch.   In both sexes, the deep pouch consists of the urogenital 'muscle sandwich' diaphragm, comprising a band of striated muscle stretching between the ischiopubic arches: the posterior part is the deep transverse perineal muscle; the anterior part, which runs around the urethra, is the sphincter urethrae muscle. The former tenses and fixes the perineal body, the latter is important in the maintenance of urinary continence. This musculature is covered both top and bottom by the two layers of fascia of the urogenital diaphragm. In both sexes, the urinary tract passes through the diaphragm as the membranous part of the urethra. In the male, the genital tract also exits through the urethra; in the female, the genital tract remains separate and passes posterior to the urethra as the vagina. In both sexes, the genital tract is lubricated by a set of glands: in the male, these bulbourethral or Cowper's glands reside in the substance of the diaphragm and thus inhabit the deep pouch; in the female, the homologous structures are the greater vestibular or Bartholin's glands which, however, lie just outside of the perineal membrane and are thus in the superficial perineal pouch, residing at the posterior extremity of the bulb of the vestibule and emptying into the distal vagina. Other structures in the deep pouch are the pudendal nerve and internal pudendal vessels which pierce and run through the diaphragm to reach their final destinations in the superficial pouch as penile or clitoridal vessels.

Be aware that the urogenital diaphragm covers the gap in the U-shaped loop of the puborectalis muscle. Through this gap, the urethra and vagina escape the pelvis to enter the diaphragm and eventually reach the outside world. In addition, the prostate with its contained prostatic portion of urethra sits immediately on the deep surface of the urogenital diaphragm.

Superficial perineal pouch.   This is where everything in the perineum superficial to the perineal membrane is found. In this space reside the external genitalia and all of its relations, the outlets of the genital and urinary tracts and a small muscle that stabilizes the perineal body.

Male superficial pouch.   This pouch is limited superficially by the deep layer of the superficial fascia of the skin. In the abdomen, this was known as Scarpa's fascia but, in the perineum, it is called Colles' fascia. Scarpa's fascia is attached to the muscle fascia of the thigh (fascia lata) caudal to the inguinal ligament. It sweeps down from the abdomen with the skin to cover the penis and scrotum, then continues to an attachment on the ischiopubic arches laterally and on the posterior margin of the urogenital diaphragm posteriorly. Anteriorly it attaches to the margin of the glans penis, thereby permitting an opening for the male genitourinary tract, the urethra.

Understand that this pouch is open up into the subcutaneous tissue of the abdomen under the Colles-Scarpa layer so that ruptures of the **penile** urethra, for example due to instrumentation or trauma, will allow the escape of urine into the superficial perineal pouch. Within the pouch, the urine is directed around the penis and scrotum by Colles' fascia but, as more urine is leaked, it then is directed up under the abdominal wall skin. The urine cannot spread any further back than the posterior margin of the urogenital diaphragm because of the fascial attachment there. Similarly, the urine cannot escape laterally into the thighs because Colles' fascia is there attached to the ischiopubic arches. There is, however, no obstruction to extravasation of urine up into the abdomen. Once in the abdomen, there is no escape of urine into the thigh as Scarpa's fascia is attached to the fascia lata just caudal to the inguinal ligament. Having this anatomical awareness will enable you to diagnose accurately the source of urine distending the penis-scrotum-abdominal wall. It does not originate from 'inside'! Urethral tears higher up allow urine to leak into the extraperitoneal tissue around the prostate and bladder.

Scrotum.   Colles' fascia and the overlying skin are pouched out to form an external housing for the testes and the trailing spermatic cords. (The female homologues of the scrotum are the labia majora.)

Penis.   Three expansile bodies (erectile or cavernous tissue) make up the penis: two lateral corpora cavernosa; and one midline corpus spongiosum. Each is bound by a tough fibrous cover –the tunica albuginea corporis cavernosa and the tunica albuginea corporis spongiosi, respectively. The three are bound together by a tough deep penile fascia, known as Buck's fascia, derived from the perineal membrane, to form the penile shaft. The shaft of the penis is held up against the symphysis by a suspensory ligament. Distally the three bodies are as one; proximally the two corpora cavernosa separate to run down the ischiopubic rami on each side and are tightly attached to these bones. Here they are termed the crura (legs) of the penis. The medial corpus spongiosum runs back in the midline to a point of attachment in the perineal body. The corpus spongiosum enlarges at its proximal end to form the bulb of the penis, and at its distal end to form the glans penis.

Each of the three bodies of erectile tissue is clothed by thin muscle in its proximal portion: the ischiocavernosus muscle over each crus, and the bulbospongiosus muscle arising from the perineal body and covering the proximal corpus spongiosum. These muscles extend up to the base of the penile shaft. Their contraction impedes the return of venous blood from the three corpora. When arterial inflow to the cavernous bodies concurrently is increased by dilation of branches of the internal pudendal artery (deep artery of the penis and the artery to the bulb) to these structures, the corpora become tightly distended with blood, thus enlarging and rendering the penis erect. Voluntary contraction of the bulbospongiosus ejects the urine remaining in the penile urethra after bladder-emptying is completed.

The superficial transverse perineal muscle runs from each side of the ischiopubic arch to stabilize the perineal body. As it overlies the posterior margin of the urogenital diaphragm, the muscle is an important surgical landmark.

The urethra, after traversing the urogenital diaphragm where it is sphinctered by the sphincter urethrae and lubricated by Cowper's glands, enters the posterior superior aspect of the bulb of the urethra and becomes the penile urethra.

Female superficial perineal pouch. The clitoris is a small edition of the male penis, but differs by having no intimate relationship to the urethra and by its construction out of two, rather than three, cavernous bodies – the corpora cavernosa clitoridis. It has entirely similar crura of erectile tissue that arise from the ischiopubic rami and meet to form a shaft with a suspensory ligament at the symphysis, then continue forwards to end in a small bulbous glans or head. Ischiocavernous muscles clothe the crura just as in the male.

As the genital tract remains separate from the urinary tract, the two exit separately in the vestibule of the vagina after passing through the urogenital diaphragm. The vestibule is the space between and surrounded by the labia minora, two skin folds homologous to penile skin in males. They surmount the female corpora spongiosa (vestibular bulbs). These erectile tissues are not joined around the urethra as in the male, but separately pass around the vagina. These cavernous bodies are clothed in thin striated muscle as in males – the bulbospongiosus muscles. Together with the ischiocavernosa, these muscles produce tumescence of the vestibular bulbs and erection of the clitoris. On their own, the bulbospongiosa act as a constrictor of the vagina.

The urethra exits in the anterior portion of the vestibule immediately after its passage through the fascia and muscles of the urogenital diaphragm.

Homologous to the Cowper's glands in males and similarly lubricating the genital tract outlet (vagina) are two 'greater vestibular' or Bartholin's glands subjacent to the vestibular bulbs. These are concealed by the bulbospongiosus muscles and empty into the distal vagina via a pair of short ducts. These glands frequently become infected and require drainage. The difference between males and females here is the location of the female glands in the superficial pouch instead of in the substance of the urogenital diaphragm (the deep pouch) as in males.

The labia majora are a second set of skin folds that are larger than the minora and situated just outside of the minora. These are filled with fat and the terminal strands of the round ligament; the whole is the homologue of the male scrotum and gubernaculum.

Pudendal nerve and internal pudendal vessels. These supply everything in the perineum (Figure 4.1) except for some anterolateral scrotal-labial skin (ilioinguinal and genitofemoral nerves) and some posterolateral scrotal-labial skin (posterior femoral cutaneous nerve from the sciatic plexus). The pudendal nerve and the internal pudendal vessels exit the true pelvis through the greater sciatic foramen under the inferior border of piriformis along with all other exiting structures except the superior gluteals, which exit superior to piriformis. However, the pudendals reenter the pelvis by winding around the sacrospinous ligament and enter the perineum by passing between the sacrospinous and sacrotuberous ligaments (lesser sciatic foramen). They come to lie on the surface of obturator internus and are distal to (caudal-superficial to or outside of) the take-off of levator from the obturator internus fascia. Thus, they run within the lateral wall of the

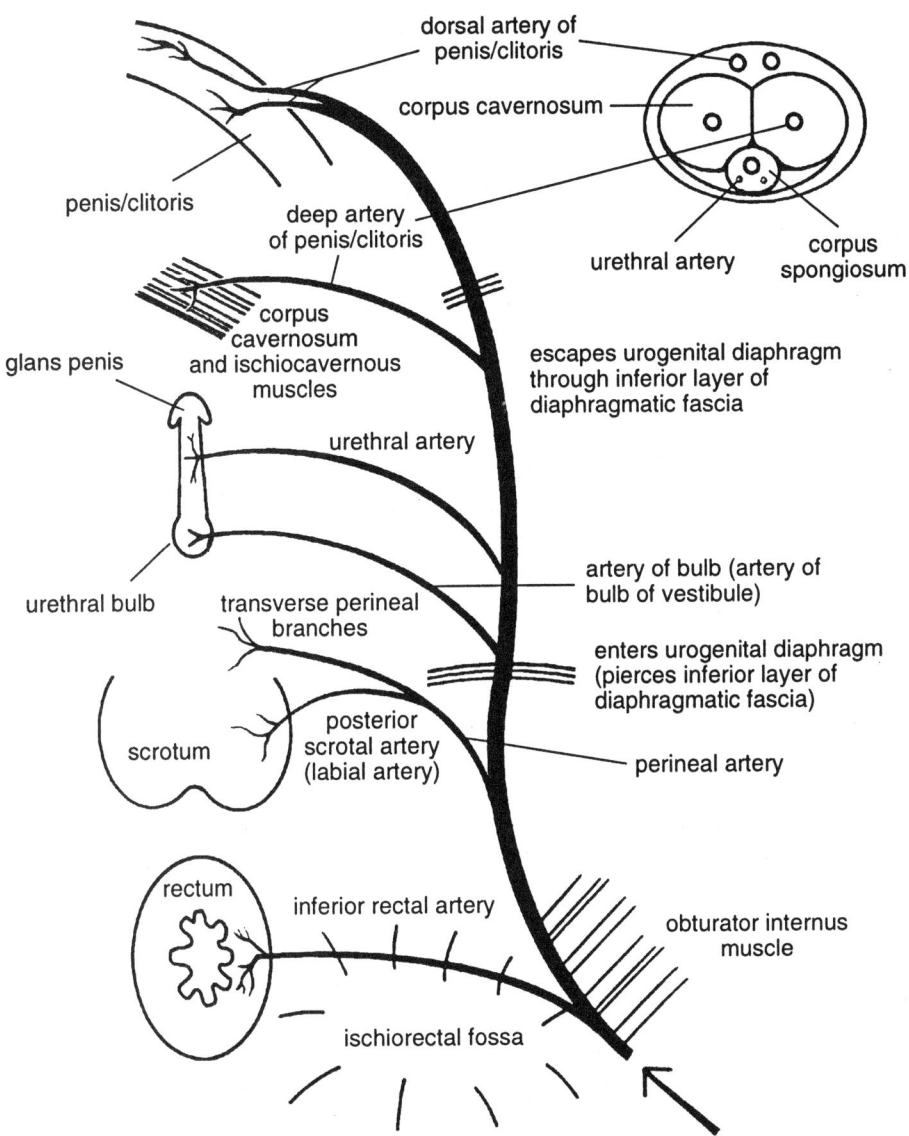

Figure 4.1 Internal pudendal artery in the perineum. (The pudendal nerve branches accompany the arteries)

ischiorectal fossa. Just before reaching the tuberosity, they give off the inferior rectal vessels and nerves supplying the anal canal area, then turn upwards to run along the ischiopubic arch, giving off a superficial supply to the skin and a deep supply to the cavernous tissue [artery to the bulb (urethral or vestibular), artery to urethra, deep artery of the penis or clitoris]. These deep branches are given off as the pudendals traverse the urogenital diaphragm. Finally, the pudendals escape the urogenital diaphragm to end as the dorsal artery and nerve of the penis-clitoris.

Pelvis and Perineum Self-Assessment

1. The true pelvis is marked by:

   A   Iliac crest.
   B   Iliopectineal line (arcuate line).
   C   Reflection of peritoneum onto the rectum.
   D   Sacroiliac joint.
   E   Obturator canal.

2. Which one of the items below passes through the lesser sciatic foramen?

   A   Piriformis muscle.
   B   Inferior gluteal vessels and nerve.
   C   Obturator externus muscle.
   D   Tendon of the obturator internus muscle.
   E   Sciatic nerve.

3. An abscess in the peritoneal rectouterine pouch might be recognized by:

   A   Rectal examination.
   B   Abdominal palpation.
   C   Chest x-ray.
   D   Stool culture.
   E   Spinal tap.

4. Regarding the rectum, one statement is correct:

   A   It is strongly supported by its lateral ligaments.
   B   Its nerve supply comes from autonomic fibers of L1–2 and the vagus.
   C   Its chief blood supply comes from the terminal branch of the inferior mesenteric artery.
   D   It passes through the urogenital diaphragm to reach the anus.
   E   It is crossed in its lower portion by the ureter.

5. Choose the incorrect ending to the following statement: The hypogastric artery is an important source of blood for the:

   A   Bladder.
   B   Uterus.
   C   Ovaries.
   D   Gluteal muscles.
   E   Anal canal.

6. Regarding the urinary tract in the pelvis, choose the one incorrect statement:

   A   The bladder is closely related posteriorly to the rectum in the male, but shares no blood supply with the rectum.
   B   Ureters are crossed anteriorly by either the uterine artery or the ductus deferens.
   C   The urinary bladder is composed of smooth muscle.
   D   The lower ureters receive blood supply from the inferior vesical arteries.
   E   The ureters terminate in the prostatic urethra.

7. Regarding the prostate, choose the one incorrect statement:

   A   Contains glands which elaborate their secretions into the bladder.
   B   Rests on the deep surface of the urogenital diaphragm.
   C   Is immediately anterior to the rectum and may be readily palpated by digital rectal examination.
   D   Ejaculatory ducts run through the prostate.
   E   Contains the most proximal portion of the urethra.

8. Regarding the female internal genitalia, choose the one incorrect statement:

   A   Sympathetic nerve supply to the uterus comes from fibers accompanying the ovarian arteries.
   B   The uterus receives important support from the transverse cervical ligament and the uterosacral ligament.
   C   The round ligament of the uterus enters the inguinal canal.
   D   The uterine tubes are contained in the broad ligament.
   E   The vagina passes through the urogenital diaphragm.

9. In cases of transection of the mid-thoracic spinal cord, choose the one incorrect statement:

   A   Sympathetic supply to the pelvic viscera is not interrupted.
   B   Parasympathetic supply to the pelvic viscera remains as these fibers come from S2–4.
   C   The micturition reflex continues to be stimulated by distention of the bladder.
   D   The reflex arc governing defecation is interrupted.
   E   Sensations of pain arising in the pelvic viscera are eliminated.

10. Regarding the levator ani muscle:

   A   This is smooth 'automatic' muscle tissue.
   B   Arises in part from the fascia of the external obturator muscle.
   C   Forms important support for the urogenital diaphragm.
   D   Closes off the rectum except during the process of defecation.
   E   Forms the lateral border of the ischiorectal fossa.

11. Regarding the perineum, choose the one incorrect statement:

    A   In both sexes includes all tissue caudal to levator ani.
    B   Includes the ischiorectal fossa.
    C   Its structures are largely supplied by the internal pudendal artery.
    D   The midpoint of the perineum (perineal body) represents the area of penetration of the urorectal septum to the skin.
    E   May be divided into urogenital and anorectal triangles by a line projected between the two ischial spines.

12. The deep perineal pouch includes:

    A   Deep and superficial transverse perineal muscles.
    B   Prostatic urethra.
    C   No erectile tissue.
    D   Bartholin's greater vestibular glands.
    E   Smooth muscle of the penile crura.

13. Choose the incorrect ending to the following statement: Extravasation of urine due to rupture of the penile urethra and its containing fascial tunica will:

    A   Fill the superficial pouch with urine.
    B   Pass up into the abdominal wall under the deep layer of the subcutaneous fascia (Scarpa-Colles).
    C   Work its way via openings in the urogenital diaphragm into the perivesical space.
    D   Not pass into the thigh from the abdominal wall.
    E   Not infiltrate perianal tissue due to the attachment of the deep layer of subcutaneous fascia to the superficial layer of the fascia of the urogenital diaphram.

14. Regarding the anal canal, choose the one incorrect statement:

    A   Is formed from the terminal hindgut and the proctodeum.
    B   Contains the pectinate line demarcating the junction of squamous and columnar epithelium.
    C   Extends from the puborectalis portion of levator ani to the anal verge skin.
    D   Proximal to the pectinate line, the venous drainage is to the inferior vena cava.
    E   The puborectalis muscle is vitally important in the maintenance of fecal continence.

15. The ischiorectal fossa:

    A   Contains the inferior rectal artery.
    B   Has the levator ani muscle as its caudal boundary.
    C   Lies lateral to the internal obturator muscle.
    D   Communicates with the abdomen via the obturator canal.
    E   If infected, must not be incised because of the likelihood of producing incontinence.

16. Regarding the external genitalia, choose the one incorrect statement:

    A The urethra in both sexes is supported by striated bulbospongiosus muscle.
    B Erectile tissue here depends on an adequate flow of blood through the internal pudendal artery for satisfactory function.
    C All external genitalia are located in the superficial perineal pouch.
    D The vagina is supported by the perineal body.
    E In both sexes, sensory innervation is supplied by the pudendal nerve and autonomic innervation by sympathetic and parasympathetic fibers from the pelvic plexus.

For each numbered item, select the homologous lettered item.

    A Bartholin's vestibular gland.
    B Round ligament of the uterus.
    C Labia majora.
    D Vestibular bulb.
    E Crus clitoris.
    F Urogenital diaphragm.

17. Corpus cavernosum penis.

18. Scrotum.

19. Corpus spongiosum.

20. Bulbourethral glands.

21. Gubernaculum.

Answers and Explanations

1. **B**

2. **D** **A**, **B** and **E** pass through the greater sciatic foramen; **C** passes through neither.

3. **A** In the adult, rectouterine pouch contents are too far from the abdominal wall to be palpable. None of the studies listed help.

4. **C** The middle rectal (hemorrhoidal) artery is so small, it is insignificant. The lateral ligaments of the rectum have no strength, but are merely limp fibrous causeways for the midrectal arteries, veins, lymphatics and autonomic nerves. Parasympathetics arise in S2–4 nerves, not the vagus. The urogenital diaphragm is confined to urinary and genital tract structures. The ureter enters the bladder without crossing the rectum.

5. **C** Ovarian blood comes from the ovarian arteries, direct branches of the abdominal aorta. The anal canal is supplied by the inferior hemorrhoidal branch of the inferior pudendal, a branch of the hypogastric. Vesical, uterine and gluteal arteries are all branches of the hypogastric.

6. **E** Ureters terminate in the bladder. Bladder and rectum are indeed close, but are separated by the urorectal septum, which forms a bloodless plane between the two (same as with rectum and vagina).

7. **A** Prostatic secretions are ducted to the prostatic urethra.

8. **A** Uterine nerve supply comes from the pelvic plexus via the transverse cervical ligament – uterine artery. The genital and urinary tracts in both sexes pass through the urogenital diaphragm.

9. **D** Theoretically at least, the reflex arcs of micturition and defecation would persist despite such cord transection. The highest level of these reflex arcs is in the L1 cord segment, which lies at around T11 vertebral level. Conscious sensation and control is lost. The reflexes would continue automatically.

10. **D** Levator ani in its puborectalis portion is the most important muscle of fecal continence. The puborectalis 'sling' closes the rectum and merges with the more caudal external anal sphincter. It is good striated muscle, arising in part from the fascia over the internal obturator muscle, but with little functional partnership with the urogenital diaphragm. It forms the superior wall of the ischiorectal fossa.

11. **E** The dividing line runs between the ischial tuberosities.

12. **C** The superficial transverse perineal muscle is in the superficial pouch as are Bartholin's glands and all genital erectile tissue. The ischiocavernous muscles of the penile crura are striated. The urethra in the deep pouch is membranous urethra.

13. **C** The key here is that the superficial pouch is bounded by the deep layer of subcutaneous fascia (Colles' fascia), which is continuous with the deep layer of superficial fascia in the abdomen (Scarpa's fascia). Thus, the superficial pouch may be thought of as a diverticulum of the abdominal subfascial subcutaneous space that

is contiguous with it. Due to the attachment of Scarpa's and Colles' fasciae to the fascia lata, ischiopubic arch and fascia of the urogenital diaphragm, there is nowhere for a filled superficial pouch to empty, but into the abdominal wall subfascial subcutaneous space. The urogenital diaphragm is not 'porous'.

14. **D** Venous drainage cranial to the pectinate line proceeds to the portal system and, caudal to the line, to the caval system. (Some consider the anal canal to extend from the dentate line to the anal verge. Functionally, the canal is considered to include the muscle of continence, the puborectalis sling.)

15. **A** Levator ani is medial and obturator internus lateral. The obturator canal lies cranial to the origin of levator from the obturator internus fascia and is thus cranial to the fossa. Abscesses in the fossa must be incised and drained promptly to prevent major sepsis. Incontinence should not result if such a procedure is performed properly.

16. **A** Bulbospongiosus in the female is the bulb of the vestibule of the vagina. It does not encase the female urethra.

17. **E**

18. **C**

19. **D**

20. **A**

21. **B**

5 Lower Extremity

It is important to realize that the lower extremity is required to support and move considerable weight – up and down, and forward and back – in the bipedal erect human. This requires big strong bones and joints, massive muscles and a huge blood supply. Small problems may result in incapacitating pain with every step or an inability to stand or move at all. We cannot deal cards with our feet nor run on our hands. The differing anatomical requirements between arms and legs is obvious.

Bones. To describe and communicate accurately, the following terms should immediately call forth clear images in the mind's eye:

Femur	Head, fovea
	Neck
	Greater trochanter
	Lesser trochanter
	Intertrochanteric crest (posterior aspect of femur)
	Intertrochanteric line (anterior aspect of femur)
	Gluteal tubercle
	Linea aspera
	Adductor tubercle
	Medial epicondyle
	Lateral epicondyle
Patella	Medial condylar facet
	Lateral condylar facet
	Apex
Tibia	Medial condyle
	Lateral condyle
	Intercondylar eminence
	Tibial tuberosity
	Tibial crest (anterior subcutaneous margin)
	Medial malleolus
Fibula	Head
	Neck
	Lateral malleolus
Talus	Trochlea, body, head
	Superior articular surface (trochlea)
	Talocalcaneal articulation, posterior
	Talocalcaneonavicular articulation(s), anterior

Calcaneus	Talar articular surfaces
	Sustentaculum
Navicular	Talar articular surface
	Tuberosity
Fifth metatarsal	Tuberosity.

Fascia. The investing fascia of the thigh (fascia lata) is thick and strong. Along the lateral thigh, it is particularly thick and strong, and is called the iliotibial tract. The tract receives the tendon of the tensor fascia lata and most of the insertion of gluteus maximus. It inserts with the biceps femoris at the head of the fibula and neighboring tibia. In the thigh, the fascia lata gives off intermuscular septa to the femur and, in the femoral triangle area, presents an opening, the fossa ovalis, which allows the superficially placed greater saphenous vein to reach the femoral vein.

Muscles of the thigh. The primary motions are considered here from the neutral position; thus, the iliopsoas muscle prime function is flexion of the thigh. Acting with the hip joint fixed, it flexes the lumbar spine with its contralateral fellow or, acting alone, it flexes the spine laterally. It can also rotate the thigh medially.

Flexion of thigh. Iliopsoas is the main actor. It extends from the transverse processes of the lumbar spine and the iliac fossa, and under the inguinal ligament to the lesser trochanter. Innervation is L2–3 from the lumbar plexus and femoral nerve. (Other lesser flexors are sartorius, tensor fascia lata, pectineus and rectus femoris.)

Extension of thigh. The major extensor is gluteus maximus, which extends from the posterior gluteal line of the ilium, edge of the sacrum and sacrotuberous ligament to the gluteal tuberosity atop the linea aspera of the femur and to the iliotibial tract of the fascia lata. Note that the great size of gluteus maximus is a distinguishing feature of the erect biped – humans. It takes power to hold the trunk upright! Innervation is from the upper roots of the sacral plexus – L5 and S1–2 via the inferior gluteal nerve. The nerve travels to the muscle with the inferior gluteal vessels, passing from the pelvis beneath the lower border of piriformis through the greater sciatic foramen. Between gluteus maximus and the greater trochanter is the **trochanteric bursa**, which protects the muscle as it glides over the trochanter. The bursa is frequently the site of inflammation and pain – trochanteric bursitis. (Lesser extensors are biceps femoris and adductor magnus acting when the thigh is medially rotated.)

Abduction of thigh. In general, the other two glutei are responsible for abduction as they run directly from the ilium to the greater trochanter. The nerve supply is from the superior gluteal and sacral plexus, L4–5 and S1. These pass to the muscles over the upper border of piriformis, along with the superior gluteal vessels, through the greater sciatic foramen.

Gluteus medius (and minimus) contraction holds the pelvis horizontal and the torso upright when the homolateral lower extremity bears weight when walking. Paralysis of the gluteus medius results in a 'falling away' of the pelvis – a downward tilt of the pelvis and torso toward the normal side with each weight-bearing step on the involved side. This results in a characteristic limping 'gluteal gait' (Trendelenburg gait).

Adduction of thigh. The adductor group extends in series from the body of the pubis, the ischiopubic arch, to the pubic tuberosity and includes (in this order) pectineus, adductor longus, gracilis, adductor brevis and adductor magnus. From their pelvic origin, they pass medially to the linea aspera and adductor tubercle (adductor magnus) of the femur. The nerve supply is the obturator from the femoral plexus except for pectineus, which receives femoral nerve branches, and the lower part of adductor magnus, which has a hamstring function and is supplied by the tibial portion of the sciatic nerve.

Lateral rotation of thigh. Obturators internus and externus, the gemelli, piriformis and quadratus femoris all run from the pelvis to the area of the greater trochanter under cover of gluteus maximus and rotate the thigh laterally. The nerve supply is from the upper sacral plexus – L4–5 and S1–2 – except for obturator externus, which perversely accepts a branch of the obturator nerve. (All of these muscles have secondary functions in abduction and adduction of the thigh. Those inserting higher on the trochanter are better abductors but poorer adductors than those inserting lower down.)

Obturator internus is an interesting muscle well worth reviewing. Its origin is mainly the obturator membrane; its right-angled course is under the ischial spine; it serves as a source of origin for a good portion of levator ani; it constitutes the lateral wall of the ischiorectal fossa; it provides a fascial tunnel for passage of the internal pudendal vessels and nerve, and a canal for the passage of the obturator nerve and vessels.

Medial rotation of thigh. Gluteus minimus and the anterior part of gluteus medius are involved here. (The tensor fascia lata, adductors and iliopsoas can help.)

Muscles of the knee. We are primarily only interested in its movement during standing and walking.

Extension of knee. The quadriceps group includes rectus femoris and the three vasti. This large anterior mass of muscle arises from the shaft of the femur (the vasti) and the ilium (rectus femoris, from the anterior inferior iliac spine and superior margin of the acetabulum). Quadriceps inserts into the patella; innervation is by branches of the femoral nerve. The patella is a sesamoid bone in the quadriceps tendon which adds leverage to the quadriceps pull as well as additional protection to the joint. Actual quadriceps insertion is into the tubercle of the tibia. The patellar ligament is in fact the continuation of the quadriceps tendon. The quadriceps tendon sends expansions to the fibrous capsule of the knee joint (patellar retinaculum) which greatly strengthen the joint. (Rectus femoris, with its origin from the ilium, can act as a hip flexor.)

Flexion of knee. The 'hamstrings' include the 'ham', the posterior mass of the thigh, and the 'strings' are the cord-like tendons of these muscles which are readily palpable and visible as the borders of the upper half of the popliteal space. There are three hamstrings: biceps femoris (two heads); semimembranosus; and semitendinosus. All spring from the ischial tuberosity (short head of biceps from the linea aspera). The 'semis' insert on the tibia (medial strings) and biceps on the fibula (lateral string). All are innervated by the sciatic tibial division except for the short head of biceps, which accepts a branch of the peroneal division.

Two other muscles must be considered in connection with the thigh, namely, sartorius and popliteus. **Sartorius** runs diagonally across the thigh from the anterior superior spine to the medial tibia, and is an important anatomical landmark for the surgeon. *Sartor* is the Latin for tailor and this muscle helps the tailor assume the cross-legged sitting position. **Popliteus** is a thin flat muscle that lies deep to the gastrocnemius heads and forms the floor of the lower popliteal fossa. It arises from the lateral femoral condyle and inserts on the posterior aspect of the proximal tibia. Innervation is from the tibial nerve. Its tendon of origin passes between the fibular collateral ligament and attachment of the lateral meniscus, thus protecting the meniscus from pathological pull when the lateral ligament is torn. On the medial side of the knee, the medial meniscus is attached to the tibial collateral ligament. Tearing of this ligament is often accompanied by tearing of the medial meniscus. The popliteal artery bifurcates at the lower margin of the popliteus muscle. Popliteus rotates the leg medially as the initial step in 'unscrewing' the locked straight knee in passing from extension to flexion.

Muscles of the ankle.   Permitting only dorsiflexion and plantar flexion, the ankle is strictly a hinge joint. Eversion and inversion occur at the talocalcaneal and talonavicular joints.

Dorsiflexion of ankle.   This involves all of the anterior compartment muscles: tibialis anterior; extensor digitorum; and extensor hallucis. Tibialis anterior inserts on the medial aspect of the tarsus (first cuneiform) and base of the first metatarsal. Its inverting action is balanced by a synergistic contraction of peroneus brevis (lateral compartment) to produce pure dorsiflexion. All of these muscles are innervated by the peroneal nerve.

Injury to the peroneal nerve paralyzes these muscles and produces a 'foot drop'. The injured patient must walk with a high stepping gait so that the toes clear the ground. The drooping foot slaps down forefoot first.

Plantar flexion of ankle.   This involves all of the posterior compartment muscles: gastrocnemius and soleus to the calcaneus via the calcaneal (Achilles) tendon; the posterior tibial behind the medial malleolus to the tuberosity of the navicular; and the two flexors' digitorum and hallucis (after their primary job of digital flexion). These posterior muscles are all innervated by the tibial nerve. (The laterally placed peroneus muscles also plantar flex, their eversion pull balanced by the inversion pull of the posterior tibial.) (The plantaris muscle, running from just above the lateral femoral condyle to calcaneus, has a pencil-thin long tendon lying between gastrocnemius and soleus. It has no functional flexor significance, but is clinically important as it often ruptures, causing sudden calf pain and temporary dysfunction. It may be used as a tendon graft.)

Muscles of eversion and inversion of the foot.   This occurs not at the ankle joint, but principally involves the articulations between calcaneus and talus posteriorly,

110

and talus-calcaneus and talus-navicular anteriorly – the so-called **subtalar joint**, also termed the **transverse tarsal joint** (see below).

Eversion (abduction and pronation) of foot. These involve the lateral compartment muscles, peroneus longus and brevis. Peroneus brevis inserts on the head of the fifth metatarsal whereas peroneus longus crosses under the sole to insert on the head of the first metatarsal. Innervation of both is from the superficial peroneal nerve. (Aiding this eversion motion is extensor digitorum longus, which produces eversion after extending the digits.)

Inversion (adduction and supination) of foot. This is largely accomplished by tibialis anterior and posterior through their insertions into the medial portions of the tarsus.

Muscles of flexion and extension of the pedal digits.

The most important digit is the hallux, the great toe, which has its own flexor and extensor muscles in the leg. A well-functioning great toe permits walking, push-off, running, etc.; without the great toe, there is marked disability. The foot has the same little muscles as in the hand, but the important function of the foot is support and ambulation, not writing or playing the piano! Be aware of these small muscles, but know their limitations.

Flexion and extension of great toe. These are achieved by the anterior compartment extensor hallucis longus, supplied by the deep peroneal nerve, and flexor hallucis longus in the deep posterior compartment, supplied by the tibial nerve.

Flexion and extension of pedal digits. No big deal! This involves flexor and extensor digitorum longi of the posterior and anterior compartments, innervated by the posterior tibial and deep peroneal nerves, respectively.

Thickened extensions of muscle fascia form a fibrous retinaculum around the ankle which binds all of these tendons as they pass from leg to foot. Under the lateral malleolus pass the peronei, and under the medial malleolus pass tibialis posterior, and flexor digitorum longus and flexor hallucis longus. Flexor hallucis is the only one of these that passes under the sustentaculum tali of the calcaneus. Synovial tendon sheaths encase all of these tendons as they pass across the ankle from leg to foot.

Sole of foot.

Plantar aponeurosis. Thick, fibrous and tough, this extends from the calcaneus to the flexor tendons of the toes and distal skin of the sole. It resists and contains the constant trauma to the sole of the foot and is an important structure.

Intrinsic muscles of the foot. These muscles deep to the plantar aponeurosis are classically described in four layers. Be aware of them, although they need not be studied in detail.

Layer 1: Abductor hallucis
Flexor digitorum brevis
Abductor digiti quinti

Layer 2: Quadratus plantae and tendons of flexor digitorum longus
Lumbricals

Layer 3: Flexor hallucis brevis and tendon of flexor hallucis longus
Adductor hallucis
Flexor digiti quinti brevis

Layer 4: Interossei and tendon of peroneus longus

Joints. Essential for movement (ambulation), they are frequently injured or diseased to the point of an inability to stand or walk. Arthritic hips cannot be placed in a sling to carry on functioning. Lower extremity joint anatomy is extremely important to any physician who wants to understand movement, and the frequent trauma and disease of the lower extremities.

Hip joint. This large strong joint has several important anatomical features. The acetabular articular surface is enlarged by a rim of fibrocartilage, the acetabular labrum. The femoral head is securely grasped. The joint cavity is large and demarcated by a tough fibrous capsule, lined by a synovial membrane, which extends laterally from the acetabular borders to the region of the intertrochanteric line. Thus, the neck of the femur lies within the joint capsule. Blood vessels supplying the neck and head of the femur enter the bone at the level of attachment of the capsule. They cannot run 'naked' across the joint space. Fractures across the neck of the femur, a frequent occurence, may cut off the blood supply to the head, resulting in necrosis of the head. Excision and joint replacement may be necessary. The small vessel which reaches the femoral head from the obturator artery via the ligament of the femoral head (ligamentum teres) is rarely sufficient to keep the head alive.

The hip joint is strongly reinforced by ligaments, the most important of which is the iliofemoral ligament (the Y-shaped ligament of Bigelow). This extends from the anterior inferior iliac spine to the capsule margin area – the intertrochanteric line. The iliofemoral ligament resists hyperextension of the hip and maintains an erect posture without muscle action. The center of gravity of the trunk passes posterior to the rotating hip joint.

Knee joint. A complex and frequently injured joint, the joint cavity is large, extending up under the quadriceps tendon and down to the articular rims of the tibial condyles. The fibrous joint capsule is reinforced by expansions from all of the muscles running from the thigh to the tibia and fibula, and by the patellar tendon. The tibial articular surfaces are enlarged by two cartilaginous semilunar rings, the medial and lateral menisci (Gk. crescent). Intrinsic ligaments include the medial (tibial) and lateral (fibular) collateral ligaments, and the anterior and posterior cruciate ligaments.

Collateral ligaments of knee. The collateral ligaments are tough fibrous band-like supports running from femur to tibia and femur to fibula on the medial and lateral aspects of the knee, respectively. They offer static resistance to any distorting force applied to the joint from either side. The medial meniscus is attached to the medial collateral ligament whereas the lateral meniscus is separated from the lateral collateral ligament by the tendon of the popliteus muscle. Thus, a tear of the medial collateral ligament caused by a force driving the knee medially will usually also tear

112

the medial meniscus. This is not so if force is applied from the other direction: the lateral meniscus may well escape injury even when the lateral collateral is torn.

Cruciate ligaments of knee. The cruciate (crossed) ligaments are named by their points of attachment to the tibia. The anterior cruciate runs from the anterior intercondylar area of tibia back up to the medial posterior aspect of the lateral femoral condyle. The posterior cruciate runs from the posterior tibia to the medial anterior aspect of the medial femoral condyle. The two cross in the middle of the joint. The cruciates resist hyperextension and hyperflexion of the knee, and prohibit any significant anterior–posterior sliding of femur on tibia.

A frequent injury involves tearing of the medial collateral ligament, the attached medial meniscus and the anterior cruciate. This 'terrible triad' is the most common knee injury incurred by football players and requires corrective surgery (with varying degrees of success!).

Ankle joint (talocrural joint). This is a hinge joint in which functional integrity depends on a tight U-shaped bony mortise which receives and holds the 'hingeing' talus. The mortise is formed by the tibiofibular syndesmosis, the strong binding between the distal ends of these two bones. The walls of the mortise are the tibial and fibular malleoli. The superior portion of the talus fits snugly within the mortise, and articulates with the articular surfaces of the distal tibial shaft and inner aspects of the two malleoli. The joint is strongly supported medially by a fan of ligaments known collectively as the **deltoid ligament**, running from the medial malleolus to navicular, calcaneus and talus itself, and laterally by similar ligaments binding the fibular malleolus to the talus and calcaneus.

An ankle **sprain** is a tearing of the ligaments of either side. A significant sprain loosens the hold of the mortise on the talus. If the ankle is not rested (crutches and splinting), and the torn ligaments are not given time to heal, an unstable ankle with weak, loose medial or lateral ligamentous support results, leading to chronic recurring and frequent ankle sprains. Suture of the damaged ligaments may be required.

Transverse tarsal joint (talocalcaneonavicular joint). The mortise and tenon ankle joint allows only flexion and extension of the foot. Inversion and eversion movements occur for the most part at the posterior articulation between the talus and calcaneus, and at the anterior articulations between the talus and calcaneus, and talus and navicular. The talus only moves as a hinge; the foot swings back and forth on the undersurface of the talus. The other small joints of the foot have minimal movement, but permit adaptation of the foot to ground irregularities as when rock climbing, etc.

Arches of the foot. The bones in the foot are arranged in longitudinal and transverse arches. These flexible arches permit adaptability to varying surfaces and provide a shock-absorbing weight-bearing action. The high medial longitudinal arch has as its keystone the head of the talus. It extends from calcaneus to talus to navicular to first cuneiform. The talar head is supported by a stout elastic plantar calcaneonavicular ligament (the 'spring' ligament) as well as the posterior tibial muscle tendon, which reaches the navicular tuberosity by running under the talus. Thus, the long arch is normally a resilient shock-absorbing weight-transmitting structure. Stretching and wear-and-tear of the support of the talar head allows the body weight to force the talus down and medially to eventually rest on the ground; the foot twists outward and the condition of *pes planus* or 'flat foot' ensues. The lateral longitudinal arch is lower and extends from the calcaneus to cuboid to the fourth and fifth metatarsals.

The transverse arch is made up of the transverse row of metatarsal heads. It is supported by the peroneus longus muscle, which crosses the sole of the foot from the lateral to the medial side and inserts on the undersurfaces of the first metatarsal and first cuneiform.

Nerves of the lower extremity.

These involve three nerves (actually four, as the sciatic is composed of two): femoral; obturator; and sciatic (tibial and peroneal).

Femoral nerve.   Lumbar plexus (L2–4).

Muscles: Anterior thigh – iliopsoas and quadriceps; flexes thigh, extends knee;
Skin: Anterior thigh (anterior femoral cutaneous, L2–3); medial leg (saphenous, L2–3); [lateral thigh (lateral femoral cutaneous from lumbar plexus, L2–3)].

Enters thigh under inguinal ligament on the surface of iliopsoas, lateral to femoral artery and not within the femoral sheath (which is for vascular elements – artery, vein, lymphatics), and breaks up into branches immediately after entry.

Obturator nerve.   Lumbar plexus (L2–4).

Muscles: Medial thigh – adductors (except for pectineus, which uses femoral nerve, and obturator externus, which uses obturator nerve but is a lateral rotator). Medial part of adductor magnus to adductor tubercle has a hamstring function and uses the hamstring nerve – the tibial nerve;
Skin: Lower medial thigh and knee [uppermost medial thigh is ilioinguinal territory (L1)].

Enters thigh under pectineus muscle after passing through obturator canal with obturator vessels. Coming out from under pectineus, it branches to supply the adductor group and skin.

Sciatic nerve (tibial and peroneal).   Sciatic plexus (L4–5 and S1–3).

Tibial nerve (L4–5 and S1–3).
Muscles: Posterior thigh muscles – hamstrings, flex knee; posterior leg muscles – ankle and toe plantar flexors; tendons behind medial malleolus – invertors: tibialis posterior, flexor hallucis longus and flexor digitorum longus; intrinsic muscles of foot – terminal plantar branches;
Skin: Posterolateral leg and lateral foot (sural nerve, S2–3; 'short saphenous nerve'); sole of foot (plantar nerves, L4–5 and S1–2); [posterior thigh (posterior femoral cutaneous, separate branch of sciatic plexus, S1–3)].

Peroneal nerve (L4–5 and S1–3).
Muscles: Lateral compartment muscles – peronei, evertors (**superficial branch peroneal nerve**); anterior compartment muscles – dorsiflexion, toe extension (**deep branch peroneal nerve**);
Skin: Lateral leg and dorsum of foot (L5 and S1) – superficial perineal, web space between great and second toes – deep peroneal.

The sciatic nerve passes out of the pelvis through the greater sciatic foramen from beneath the lower border of piriformis. It appears in the thigh out from under the mid-lower border of gluteus maximus. The nerve proceeds down the posterior thigh to the lower half between adductor magnus and the long head of biceps femoris, where it (usually) divides into its two divisions – the tibial and peroneal. The **tibial nerve** continues straight down across the midpopliteal space between the two heads of gastrocnemius and under the soleus to end at the medial side of the Achilles tendon, dividing into its terminal medial and lateral plantar branches. The plantars pass down around the medial malleolus and spread out over the sole of the foot. The **peroneal nerve** angles towards the lateral side of the popliteal fossa along the medial border of biceps femoris. It curls around the neck of the fibula, enters the lateral (peroneal) muscle compartment and divides into the **superficial peroneal** lateral compartment nerve and the **deep peroneal** anterior compartment nerve. Branches to the compartment muscles are as described above.

Arteries of the lower extremity.

Recognize the importance of a strong vascular supply to the lower extremities. Large muscles doing heavy work have large appetites! Specialists in vascular disease spend much of their time and expertise dealing with ischemic cramping malfunction of the lower limb muscles, and warding off necrosis of lower extremity tissues. Such problems rarely affect the upper extremities. Blood is delivered primarily through the femoral artery. The obturator peters out quickly in the proximal adductor area. Its interesting but ineffectual acetabular branch running in the ligamentum teres of the femur to the femoral head has already been noted (see above description of femur). The arteries considered here are the profunda artery of the thigh, femoral artery in the thigh, the popliteal artery of the knee, the anterior and posterior tibial arteries of the leg, and the dorsalis pedis and plantar arteries of the foot. The femoral artery in the femoral triangle is discussed separately in a description of the triangle.

Profunda femoris (deep femoral artery). This is the major blood supply of the thigh. It branches off the femoral in the femoral triangle and immediately gives off its main branches, the medial and lateral circumflex arteries. As the names suggest, these arteries go around the upper femoral area medially and laterally to supply the bone and muscles in that region. As the profunda proceeds down the thigh, it gives off anterior muscle branches and perforating branches to the posterior muscles.

The circumflex vessels supply the frequently fractured femoral neck and head. They enter the neck at the site of attachment of the joint capsule at the intertrochanteric level. Fractures medial to the intertrochanteric level tear vessels bound for the head and neck of the bone and often result in necrosis of the head as the tiny obturator branch to the head through the ligamentum teres is insufficient to keep the head alive.

The circumflex vessels provide the 'crossbar' of the so-called cruciate anastomosis, which refers to the collateral circulation that refills the femoral after injury to the common femoral artery. The main source of collateral supply is the inferior gluteal artery anastomosing with the circumflex vessels.

The terminal descending branch of the lateral circumflex on the lateral side of the lower thigh anastomoses with branches (geniculates) of the popliteal artery and helps with refill of the leg arteries as necessary.

Femoral artery in thigh. The femoral artery passes down the thigh uneventfully after giving off the profunda. It enters the **adductor canal** at the apex of the femoral triangle and passes down the canal to an opening in the adductor magnus tendon called the **adductor hiatus**, an opening in the adductor magnus tendon allowing passage of femoral vessels to the popliteal space. Thereafter, the artery is the popliteal artery in the popliteal fossa. The adductor canal begins between the quadriceps and adductor groups, is covered by sartorius and ends at the adductor hiatus. This area is frequently the site of occlusive disorder of the femoral artery. Just before passing through the hiatus, a **descending genicular artery** (highest genicular or anastomoticus magnus artery) branches off the femoral and runs down the medial lower thigh. This is an important anastomotic refilling vessel in cases of lower femoral or popliteal occlusion.

Circulation around the knee. The popliteal region is a site of frequent arterial disorder perhaps due to the incessant bending of the knee and its popliteal artery. An obstructed popliteal artery may be refilled distally through its three sets of medial and lateral geniculate arteries connected to the descending branch of the lateral circumflex laterally, and through the descending genicular of the femoral medially.

Popliteal artery. After traversing the midline depths of the popliteal fossa, and having given off its paired superior, middle and inferior geniculates, the popliteal artery bifurcates at the lower border of popliteus into its terminal anterior and posterior tibial branches. The **anterior tibial artery** enters the anterior compartment of the leg just above the interosseous membrane between tibia and fibula, and continues down the anterior compartment to terminate at the ankle as the dorsalis pedis artery. The **posterior tibial artery** runs in the posterior compartment of the leg along with the tibial nerve in the gap between the deep and superficial portions of this compartment. Its peroneal branch is given off high in the leg and runs down the posterior compartment behind the fibula. (The peroneal artery itself neither reaches nor runs in the lateral compartment 'as it should'.) Posterior to the medial malleolus, the posterior tibial artery divides into its terminal medial and lateral plantar arteries.

Palpation of arterial pulsation affords the examiner an opportunity to evaluate arterial delivery to the lower extremity at various levels. The common femoral artery is superficial and readily palpable in the femoral triangle. The popliteal pulse is felt in the posterior midline of the popliteal space. The dorsalis pedis continuation of the anterior tibial over the dorsum of the foot lies just lateral to the extensor tendon of the great toe and is easily felt. The posterior tibial artery is superficial and easily palpable as it curls under the medial malleolus on its way to the plantar aspect of the foot.

Veins of the lower extremity. The veins of the lower limb are responsible for much human distress. The anatomical basis of these problems is straightforward. The main reason for the greater prevalence and significance of venous troubles in the lower limbs is the human necessity to stand upright and move about on two feet rather than four. More blood must move in and out of the lower limbs. Gravity also poses more of a problem here than in the arms.

Lower limb veins are found in two 'sets': superficial saphenous and deep femoral. The two systems are connected by valved communicating veins which normally permit blood to flow only from superficial to deep areas.

The deep femoral veins lie with the corresponding femoral arteries and branches. The saphenous veins are subcutaneous and in two parts: the greater and lesser saphenous systems. The **greater saphenous vein** runs along the medial aspect of the lower limb from foot to groin. It is consistently found at the ankle one fingerbreadth above and one fingerbreadth medial to the medial malleolus – a favorite site for a venous cut-down procedure. The vein terminates in the common femoral vein by penetrating the fascia lata through a gap termed the **fossa ovalis** in the upper femoral triangle. The **lesser saphenous vein** runs from the foot up the posterolateral leg to the popliteal vein in the popliteal fossa. It passes posterior to the lateral malleolus.

Competent venous valves are essential for the return of blood from the lower extremity to the heart when in an erect position. Leg and thigh muscles squeeze ('pump') the blood upwards in the deep system from valve to valve. *Vis a tergo* from cardiac activity is not enough to push blood upwards against the force of gravity. As the deep system is emptied, blood is 'sucked' from the valved superficial veins through the valved communicating veins to the deep system. Because of congenital defects in valve formation, the occupational need to stand for long periods without much leg muscle activity, or obstructions to flow such as a large pregnant uterus, the poorly supported (no surrounding muscle) saphenous system becomes distended and the valves of the superficial system are, or become, functionally incompetent. The saphenous system may then become a stagnant ever-increasing pool of blood, producing distended tortuous, visible, 'varicose' veins (*varus* = L. crooked). Serum leaks from the stagnant veins, thus causing the ankles to swell, and difficulties with nutrition of the skin of the leg and ankle ensue.

The deep system may also become incompetent as a complication of stasis due to, for example, long journeys by private or public transport, pregnancy, bedrest or prolonged movement-free standing (long operations), when clotting of the femoral system may supervene. The valves then become obliterated and, even if recanalization occurs, the valves are not repaired and serious difficulties with return circulation will occur (aching, swelling, skin ulceration). The inflammation associated with the clotting may well involve the lymphatics, with an even greater amount of swelling and fluid loss into the tissues of the leg. This chronic situation leads to devastating changes in the tissues of the leg, such as ulceration, scarring and discoloration, and considerable disability.

Lymphatics of the lower extremity.   Superficial and deep lymphatics accompany the superficial and deep systems of veins. A few lymph nodes are found in the popliteal fossa, but the majority of nodes lie in the groin around the greater saphenous termination. These include the superficial inguinal lymph nodes, which receive lymph from the skin of the lower extremity and lie superficial to the fascia lata. Deep inguinal nodes, lying deep to the fascia lata, are associated with the lymphatics running with the femoral vessels and draining the deep tissues of the lower extremity. Note that lymph from the skin of the lower abdomen and buttock, and external genitalia and perineum, also drains into lymph nodes in the superficial inguinal group in the groin. The superficial nodes drain into the deep inguinal nodes and thence to lymphatic vessels and nodes along the external iliac vessels.

The femoral triangle. Within this important anatomical area are found (and palpated) the arterial, venous and lymphatic supplies of the entire lower extremity. In addition, this is where the femoral nerve to the anterior thigh skin and muscle group (quadriceps) enters the limb to join these vascular elements. The artery, vein and lymphatics are enclosed and protected on entering the triangle by the fibrous **femoral sheath**, derived from abdominal transversalis fascia in front and the fascia covering iliopsoas behind. The femoral nerve is outside of and lateral to the sheath. From lateral to medial, the structures lie as they run from under the inguinal ligament: nerve, artery, vein and 'empty space'. The latter transmits the lymphatics and usually houses a lymph node. The V-shaped floor of the triangle is formed laterally by iliopsoas and medially by pectineus. The lateral and medial borders are sartorius and adductor longus, respectively. The roof is the fascia lata, pierced here by the fossa ovalis and the greater saphenous vein. At the cranialmost part of the roof lies the inguinal ligament. The triangle terminates distally where sartorius and adductor longus cross. This forms the beginning of the adductor canal, which is covered by sartorius. The profunda branch of the femoral artery is given off in the triangle as are the circumflex branches of the profunda. Although the nerve to the adductor, the obturator nerve, is technically not part of the femoral triangle, it lies just deep to its floor under the pectineus muscle. Be aware that the acetabulum and hip joint lie directly beneath the midportion of the femoral triangle, and that only the triangle floor (pectineus and iliopsoas) separate triangle content from the joint.

The 'empty space' is the **femoral canal** lymphatic channel. The canal is approximately 1 cm long and ends under the inguinal ligament as the **femoral ring**. The ring is bounded laterally by the femoral vein, medially by the **lacunar ligament**, in front by the inguinal ligament, and behind by the **pectineal ligament** (Cooper's), and pectineus and its fascia. The lacunar ligament is the strong fibrous arcuate connection reflected from the inguinal ligament onto the pectineal ligament along the horizontal ramus of the pubis. Its take-off point is situated immediately in front of the inguinal ligament insertion on the pubic tubercle.

The anatomical positions and relations of all of these structures are clinically extremely important (for example, for arterial palpation, arterial and venous puncture, femoral hernia, enlarged nodes, or arterial and venous surgical approaches).

Femoral hernia. This is a hernia through the 'weak spot' created by the femoral ring and canal. The femoral hernia is dangerous as it frequently strangulates the herniated intestine. This is due to the unyielding nature of the primarily ligamentous margins of the femoral ring – they do not stretch! Another hazard associated with this hernia and its operative repair is the frequent presence of an anomalous or accessory obturator artery arising from the distal external iliac to swing around either the medial or lateral border of the ring to reach the underlying obturator canal. Care must be taken not to injure this vessel in the process of femoral hernia repair; the result is otherwise 'hidden' bleeding into the extraperitoneal tissues.

Be aware that the level of the femoral ring is separated from that of the inguinal canal only by the inguinal ligament. It is sometimes difficult to differentiate inguinal from femoral hernias. Remember that the inguinal ligament inserts on the pubic tubercle. Thus, inguinal hernias are cranial to the level of the tubercle and femoral hernias are caudal.

Popliteal fossa. This busy diamond-shaped space behind the knee, often visited surgically, frequently offers diagnostic clues to the anatomically knowledgeable examiner. It is a common site of arterial vascular disease and an area worth knowing well!

The upper borders are formed by the diverging hamstrings: biceps femoris on the lateral side; semimembranosus and semitendinosus medially. The inferior borders are the heads of gastrocnemius. All of these borders are readily palpable. The fossa is covered by the fascia lata and its floor is the lower end of the femur, with the capsule of the knee joint posteriorly, and the proximal tibia covered by the popliteus muscle. Important structures, from the outside in, are the sciatic nerve and the popliteal vessels.

The **sciatic nerve** divides into its two divisions at the top of the fossa: the **tibial nerve** runs straight down the middle of the fossa and disappears under the tendinous arch of the soleus origin to continue down the leg between the superficial and deep parts of the posterior leg compartment; the **peroneal nerve** veers laterally at the top of the space and hugs the medial border of the biceps tendon to end at the neck of the fibula.

Popliteal vessels lie deep to the tibial nerve. The popliteal artery and vein enter the space through the adductor hiatus with the vein superficial to artery (although the arterial pulse is readily felt). These rest on the fossa floor. Superior, middle and inferior geniculate vessels are given off and received during the vertical passage of the vessel down the midline of the fossa. After crossing popliteus at the lower aspect of the fossa, the popliteals divide into anterior and posterior tibial branches. The **anterior tibials** disappear anteriorly through the gap above the interosseous membrane of the leg to run down the anterior compartment of the leg in company with the deep peroneal nerve. The **posterior tibials** disappear from the fossa under the tendinous arch of the soleus as did the tibial nerve. Both artery and nerve run down the posterior compartment of the leg between its superficial and deep parts. The **lesser saphenous vein** pierces the fascia lata over the popliteal fossa to enter the popliteal vein in its midportion.

Compartments of the lower extremity. Knowledge of these anatomical compartments and especially their cross-sectional anatomy will help you to organize the anatomy of the lower extremity. Compartmental anatomy enables understanding of the signs and symptoms of the lower extremity disorders that are clinically important. The compartments are composed of muscle groups, and are confined and separated from each other by indistensible fibrous septa. It is important to recognize the signs and symptoms of intracompartmental swelling and increased pressure (ranging from pain to numbness, discoloration and loss of muscle function). Such swelling and increased pressure in the compartment is often found in cases of injury to bone or muscle with the inevitable accompanying bleeding and inflammatory tissue swelling. Timely incision of the confining septum may save an extremity that might otherwise be destroyed by strangulation of the compartment tissues.

Thigh compartments.   There are three:

  Anterior compartment: Quadriceps (femoral nerve territory)
  Medial compartment: Adductors (obturator nerve territory)
  Posterior compartment: Hamstrings (tibial nerve territory).

119

Leg compartments.   There are three:

Anterior compartment: Dorsiflexors of foot and extensors of toes (deep peroneal nerve and anterior tibial artery);

Lateral compartment: Evertors and plantar flexors of foot (superficial peroneal nerve), branches of peroneal artery (peroneal artery itself runs in the posterior compartment as a branch of the posterior tibial);

Posterior compartment: Superficial and deep:

Superficial – 'calf muscles' gastrocnemius and soleus, plantar flexion, tibial nerve, posterior tibial artery;

Deep – invertors, plantar flexors and digit flexors, tibial nerve, posterior tibial and peroneal arteries.

Review Exercises

Sketch in the following on the appropriate diagrams (on page 123):

Mid-thigh

1. Three muscle compartments: Show intercompartmental septa and

    Anterior compartment
    - 1a. Rectus femoris
    - 1b. Three vasti
    - 1c. Sartorius

    Medial compartment
    - 1d. Adductors longus and magnus
    - 1e. Gracilis

    Posterior compartment
    - 1f. Semimembranosus
    - 1g. Semitendinosus
    - 1h. Biceps femoris.

2. Position of femoral vessels (femoral canal).
3. Profunda artery.
4. Sciatic nerve.
5. Greater saphenous vein.

Mid-calf

1. Three muscle compartments: Show intercompartmental septa and

    Anterior compartment
    - 1a. Tibialis anterior
    - 1b. Extensor hallucis longus
    - 1c. Extensor digitorum longus

    Lateral compartment
    - 1d. Peroneus longus
    - 1e. Peroneus brevis

    Deep posterior compartment
    - 1f. Tibialis posterior
    - 1g. Flexor hallucis longus
    - 1h. Flexor digitorum longus

    Superficial posterior compartment
    - 1i. Soleus
    - 1j. Gastrocnemius.

2. Superficial peroneal nerve.
3. Deep peroneal nerve.

4. Tibial nerve.
5. Anterior tibial vessels.
6. Posterior tibial vessels.
7. Peroneal vessels.
8. Greater saphenous vein.
9. Lesser saphenous vein.

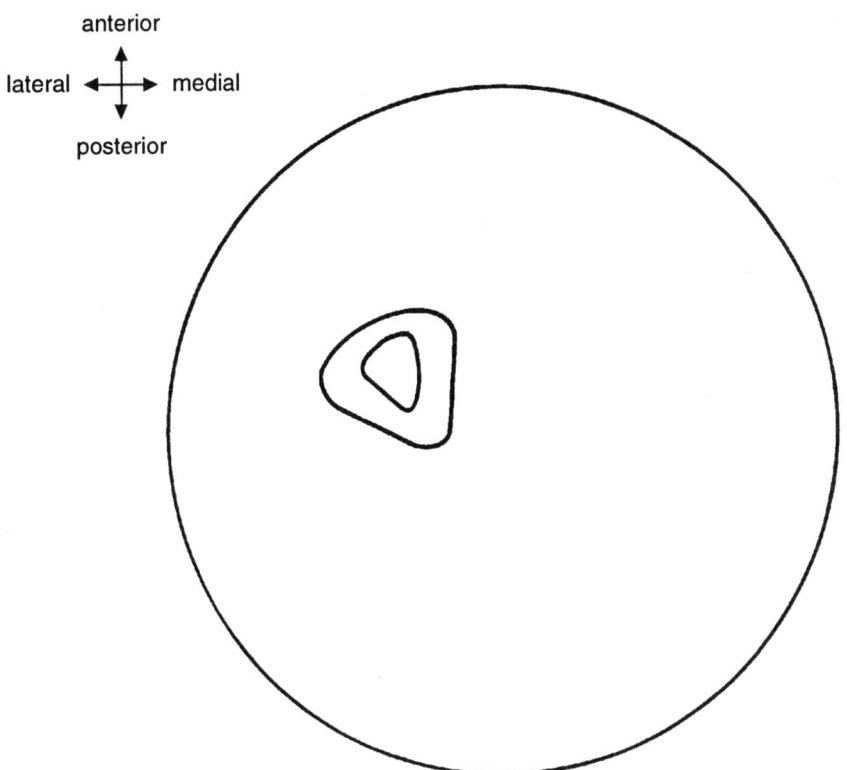

Mid-thigh

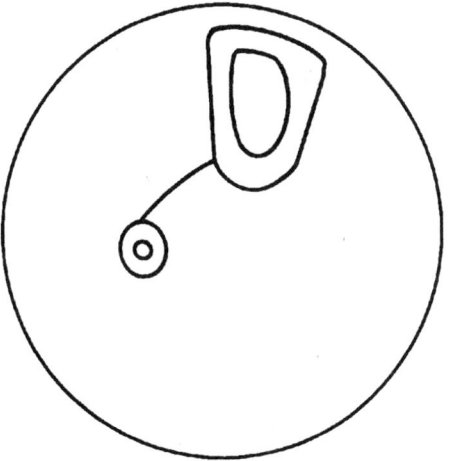

Mid-calf

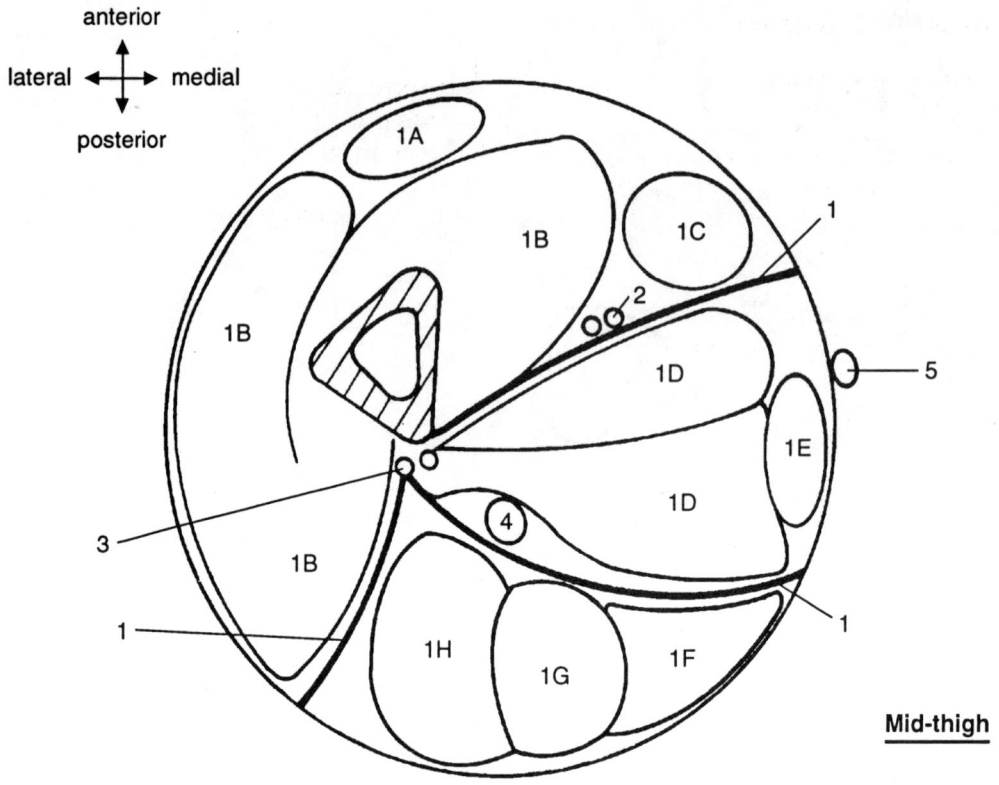

anterior

lateral ←→ medial

posterior

Mid-thigh

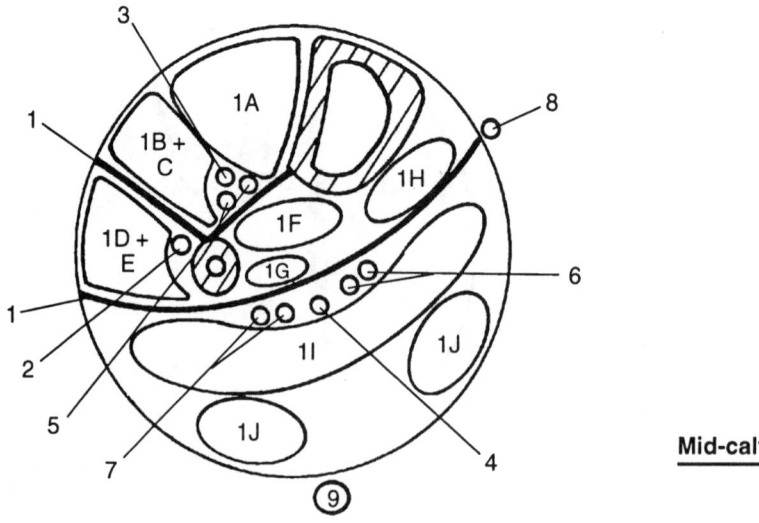

Mid-calf

Sketch the femoral arterial tree showing:

Femoral artery
Profunda femoris
Circumflex femoral arteries
Genicular branches of femoral and
    profunda

Popliteal
Genicular branches, popliteal
Bifurcation, popliteal, relation to
    popliteus muscle

Anterior tibial
Dorsalis pedis
Posterior tibial
Peroneal
Medial and lateral plantars

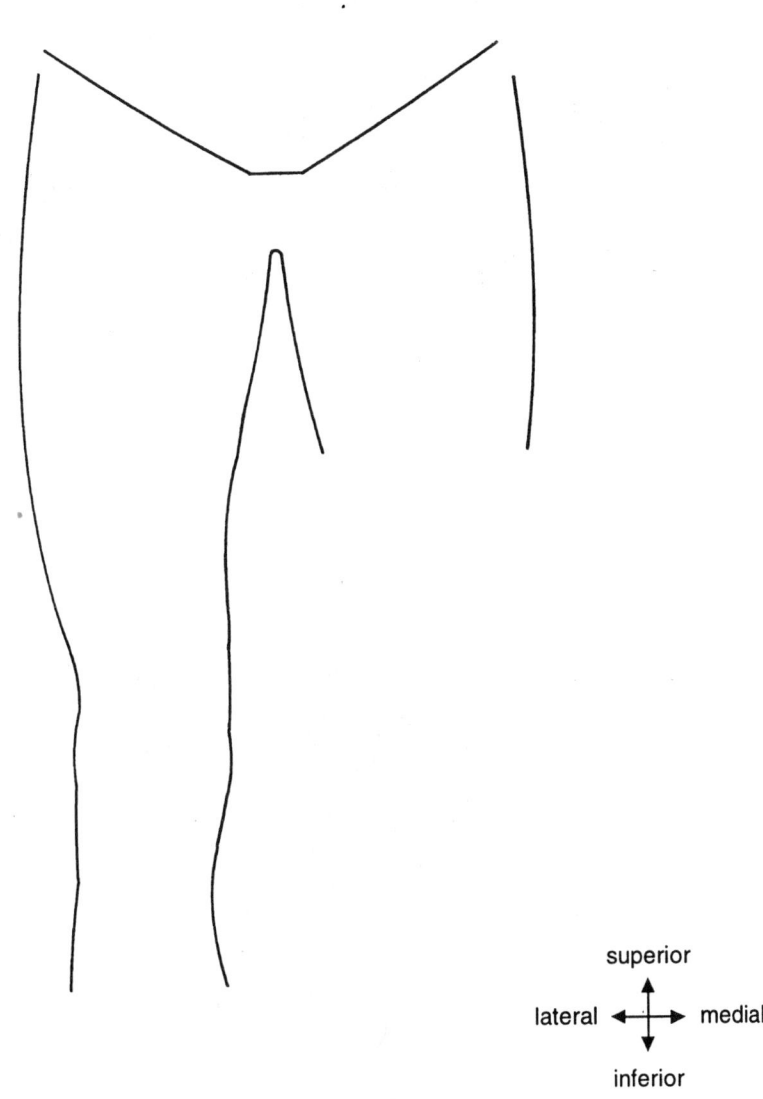

superior

lateral ← → medial

inferior

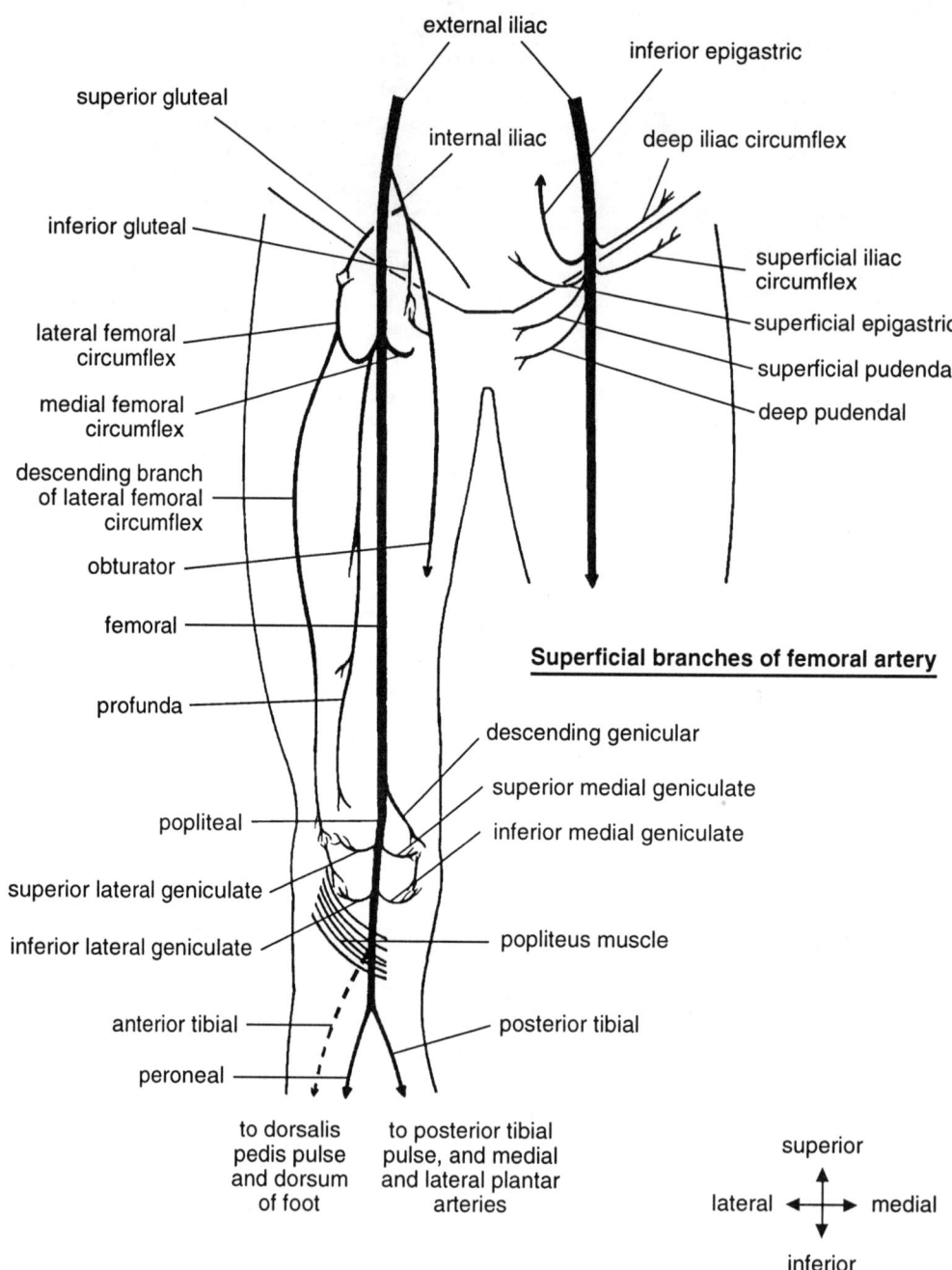

Femoral artery. Note anastomoses around the hip and knee

Femoral triangle. Sketch the inguinal ligament; muscular sides and floor; femoral artery, vein and nerve; and greater saphenous vein.

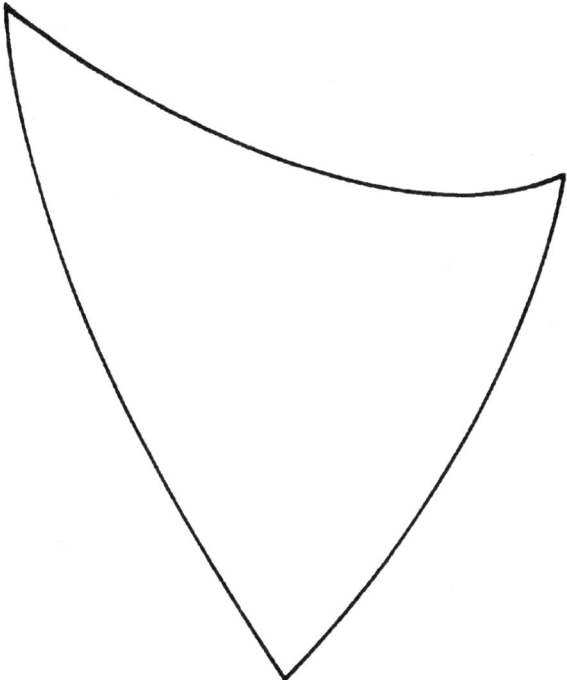

Femoral triangle 'inlet'. Sketch the femoral nerve, artery, vein and canal; inguinal, femoral and iliopectineal ligaments; femoral sheath, and positions of acetabulum and obturator foramen.

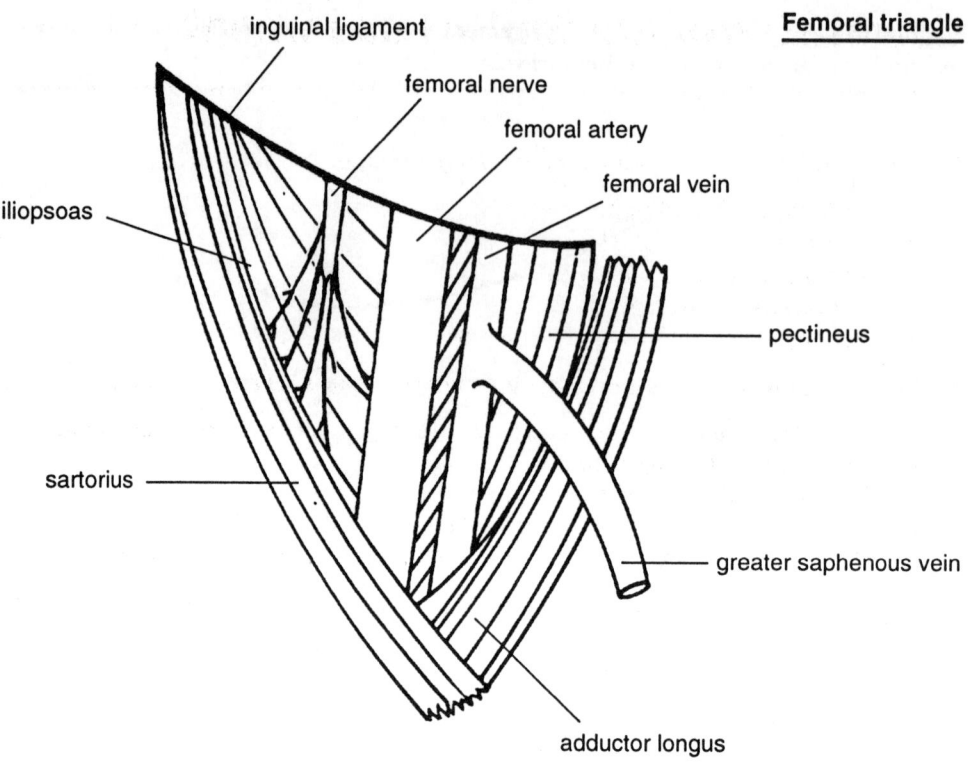

Femoral triangle

inguinal ligament

femoral nerve

femoral artery

femoral vein

iliopsoas

pectineus

sartorius

greater saphenous vein

adductor longus

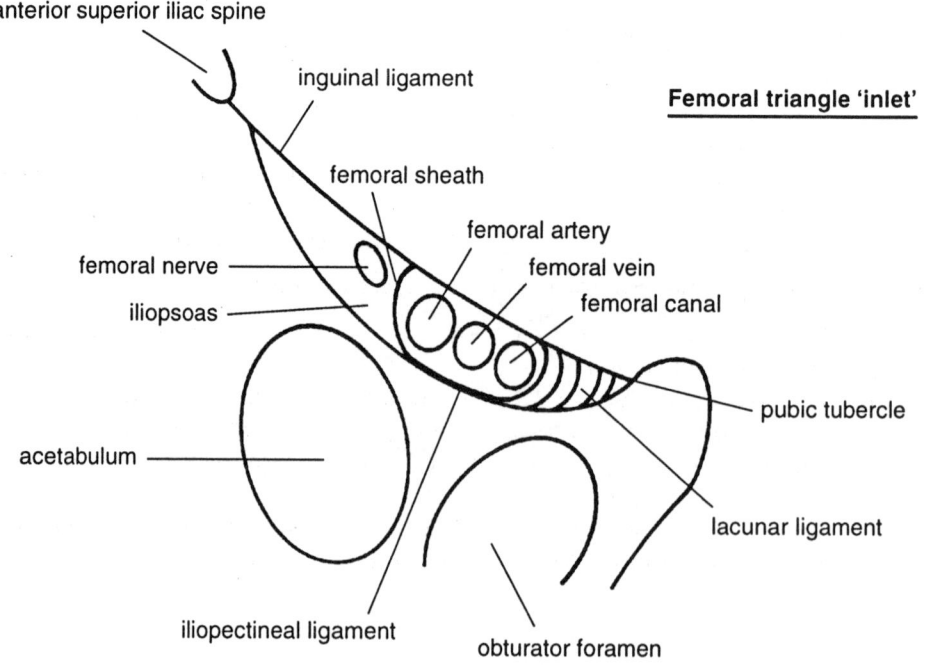

anterior superior iliac spine

inguinal ligament

Femoral triangle 'inlet'

femoral sheath

femoral artery

femoral vein

femoral canal

femoral nerve

iliopsoas

pubic tubercle

acetabulum

lacunar ligament

iliopectineal ligament

obturator foramen

Lower Extremity Self-Assessment

1. Symptoms from injury to the femoral nerve include:

   A   Difficulty in extending the leg.
   B   Difficulty in flexing the knee.
   C   Numbness of the posterior thigh.
   D   Inability to curl the toes.
   E   Atrophy of the gluteus medius muscle.

2. Regarding the adductor muscle group, choose the one correct statement:

   A   Innervation of the adductors is by the obturator, femoral and tibial nerves.
   B   Includes the gluteus maximus.
   C   Paralysis results in foot drop.
   D   In general insert on the tibial medial condyle.
   E   The popliteus muscle may be considered one of the adductor group.

3. Choose the one incorrect ending to the following statement: An injury involving the anterior muscle compartment of the leg could result in:

   A   Inability to evert the foot.
   B   Foot drop.
   C   Necrosis of the tibialis anterior muscle.
   D   Loss of the posterior tibial pulse.
   E   Weakness in extension of the great toe.

4. Fracture of the neck of the femur is a serious injury because:

   A   The femoral nerve is frequently torn.
   B   Even when healed, the patient cannot walk without a painful limp.
   C   Dislocation of the head of the femur is a frequent concomitant.
   D   Necrosis of the head of the femur may ensue due to interruption of its blood supply.
   E   Venous return from the lower limb is usually compromised.

5. Regarding the knee, choose the one incorrect statement:

   A   Tearing of the anterior cruciate ligament permits unusual anterior movement of the tibia on the femur.
   B   The tibial collateral ligament is firmly attached to the medial meniscus.
   C   The medial and lateral menisci afford great strength to the joint.
   D   The tough joint capsule is reinforced by tendinous expansions from muscles running from thigh and pelvis to tibia.
   E   The patella may be considered a sesamoid bone in the quadriceps tendon giving mechanical advantage to quadriceps function.

6. On exposing a femoral hernia, the operator will find:

   A   A peritoneal sac protruding through the direct inguinal triangle.
   B   A hernia resting on the origin of the pectineus muscle.
   C   A hernia pressing on the obturator nerve.
   D   Blockage of the femoral vein.
   E   The medial border of the hernia is the pectineal ligament of Cooper.

7. Regarding the superficial venous system of the lower extremity, choose the one incorrect statement:

   A   In the erect position, the superfical veins are emptied through valved communicating veins.
   B   Primary varicose veins are due to valvular incompetence, either congenital or acquired.
   C   In the erect position, the valved deep venous system is emptied by the squeezing of the surrounding musculature.
   D   The greater saphenous vein joins the femoral vein in the femoral triangle.
   E   Clotting in the deep system due to inflammation or prolonged stasis will not produce deleterious effects as long as the superficial system is intact.

8. In the presence of occlusion of the popliteal artery:

   A   Blood will be delivered to the leg via the obturator artery.
   B   Collateral circulation is available via the descending genicular branch of the femoral.
   C   Popliteal geniculate branches will not be available.
   D   The connection of the profunda femoris artery and the popliteal is of no use.
   E   The leg will certainly become gangrenous and require amputation.

9. Choose the one incorrect statement: In the popliteal fossa,

   A   The lateral border is biceps femoris and the lateral head of gastrocnemius.
   B   The tibial nerve runs down the midline.
   C   The popliteal artery is deeply situated, lying on the posterior aspect of the femur.
   D   The peroneal nerve runs along the medial border, which consists of semi-membranosus and semitendinosus.
   E   The popliteal vein is deep-lying, but superficial to the popliteal artery.

10. Regarding the talocrural (ankle) joint, choose the one incorrect statement:

   A   The tibia and fibula are tightly bound together (syndesmosis) forming a rigid receptacle for the trochlea of the talus.
   B   The talar trochlea presents cartilage-covered articular surfaces superiorly, laterally and medially.
   C   Strong medial (deltoid) and lateral ligaments support the joint.
   D   The tibial and fibular malleoli form the sides of the joint.
   E   Inversion and eversion of the foot take place here.

11. Regarding the foot, choose the one incorrect statement:

    A   The greater saphenous vein may be isolated superior and medial to the lateral malleolus.
    B   The head of the talus is supported by the posterior tibial muscle and the plantar calcaneonavicular ('spring') ligament.
    C   The talocalcaneal joint posteriorly and the talocalcaneonavicular joints anteriorly provide the possibility of motion generated by pull of the peroneal muscles.
    D   Sensation on the sole of the foot is carried by the plantar branches of the posterior tibial nerve.
    E   The high point of the longitudinal arch is the talus.

12. Regarding the femoral artery, choose the one incorrect statement:

    A   Is enclosed within the fascia of the femoral ring.
    B   Supplies blood to the thigh through its profunda branch.
    C   In the adductor canal is covered by the sartorius muscle.
    D   Becomes the popliteal artery after passing through the adductor hiatus of the adductor magnus muscle.
    E   If occluded in the femoral triangle, may be refilled via circumflex branches of the profunda connecting with the inferior gluteal and obturator arteries.

For each numbered item, select the closest related lettered item.

    A   Gluteus medius
    B   Hamstring muscles
    C   Iliopsoas
    D   Pectineus
    E   Peroneus longus
    F   Adductor longus

13. Plantar flexor.

14. Abduction thigh.

15. Flexion thigh.

16. Eversion foot.

17. Flexion knee.

Answers and Explanations

1. **A** Leg extension is quadriceps function, all innervated by femoral nerve. Flexion of the knee requires the tibial portion of the sciatic innervation of the hamstrings. Posterior thigh is posterior femoral cutaneous nerve territory. Toe flexors are animated by the tibial nerve. Gluteus medius is innervated by the superior gluteal nerve.

2. **A** Pectineus is innervated by the femoral nerve and lower portion of adductor magnus by the tibial nerve (hamstring function of adductor magnus); all others of adductor group are served by obturator nerve. Origins are from the pubis body and ischiopubic arch to the ischial tuberosity. Insertion is on the femoral linea aspera and adductor tubercle. Popliteus and gluteus maximus do not adduct. Foot drop is due to loss of anterior compartment muscle function.

3. **A** Eversion is by lateral compartment muscles. Anterior compartment post-traumatic swelling can choke off blood flow through the anterior tibial artery, resulting in loss of posterior tibial pulse, and necrosis of the deep peroneal nerve and anterior compartment muscles (including tibialis anterior and extensor hallucis longus). Their loss results in foot drop and loss of great toe extension. (Extensor hallucis brevis is also innervated by the deep peroneal nerve.)

4. **D** Blood supply of the femoral head comes from branches of the circumflex arteries of the profunda. These run to the head through the bony neck of the femur, entering the bone at the site of attachment of the hip joint capsule. Fractures through the neck may tear these vessels. Deprived of blood, the femoral head may suffer ischemic necrosis. The small obturator arterial branch to the head is too small to be of help. Otherwise, treatment of the fractured neck is satisfactory and free of the listed complications.

5. **C** The meniscal joint cartilages deepen the joint space, increase stability and guide some movements. They do not increase joint strength.

6. **B** The femoral ring through which a femoral hernia passes is bordered by the inguinal (superiorly), lacunar (medially) and pectinate (inferiorly) ligaments. The pectineus muscle originates in the pectinate ligament and is thus immediately posterior to the hernia sac. The obturator nerve is deep to and protected by the superior ramus of the pubis. The femoral vein, although the lateral neighbor of a femoral hernia, is not affected by the small neck of the hernia sac.

7. **E** The superficial system depends on an intact deep system for emptying except when reclining. Destruction of deep and communicating veins and their valves, as in thrombosis, produces distinctly deleterious effects on tissue of the lower extremity due to chronic venous stasis.

8. **B** Obstruction of the popliteal artery is common especially in the area of the adductor hiatus. Collateral circulation via femoral and profunda branches, genicular popliteal branches and tibial recurrent vessels is available to refill the popliteal distal to occlusion.

Vascular surgical maneuvers bypassing the obstruction are widely available. Gangrene and amputation can be avoided in the vast majority of cases.

9. **D** The peroneal nerve runs along the lateral upper border of the popliteal space, the biceps femoris muscle.

10. **E** Inversion and eversion take place in the 'subtalar joint' between talus and calcaneus posteriorly, and in the talocalcaneonavicular joints between talus, calcaneus and navicular anteriorly (see following question). The ankle joint permits only hinge motion.

11. **A** The greater saphenous vein runs one fingerbreadth above and one fingerbreadth medial to the medial malleolus.

12. **A** The femoral ring forms the passageway for lymphatics of the lower extremity. The artery is well lateral.

13. **E**

14. **A**

15. **C**

16. **E**

17. **B**

 **Upper Extremity**

In this trip through the upper extremity, much can be seen or felt and therefore studied to good advantage, using yourself or a partner as a model.

Bones. Each of the following should conjure up a definite mind's-eye picture.

Clavicle
    Sternal extremity
    Body
    Acromial extremity

Scapula
    Coracoid process
    Spine
    Supraspinous and infraspinous fossae
    Acromion
    Scapular notch
    Glenoid cavity
    Supra- and infraglenoid tubercles
    Superior, inferior and lateral angles

Humerus
    Head
        Greater tubercle and crest
        Lesser tubercle and crest
        Intertubercular (bicipital) groove
        Anatomical neck (site of attachment of joint capsule at
            borders of articular surface)
        Surgical neck (where head narrows down to shaft)

Shaft
    Deltoid tuberosity
    Radial sulcus

Condyle
    Medial and lateral epicondyles
    Capitulum
    Trochlea
    Olecranon fossa
    Coronoid fossa

Ulna
> Olecranon
> Coronoid process
> Trochlear notch
> Tuberosity (brachialis muscle insertion)
> Radial notch
> Supinator crest
> Head (distal end of ulna)
> Styloid process

Radius
> Head
>> Humeral articular surface
>> Ulnar articular surface
> Neck
> Tuberosity (biceps insertion)
> Styloid process
> Ulnar notch (distal end of radius)
> Carpal articular surface

Pectoral girdle. Understand that the arm is suspended by the pectoral girdle, which consists of the clavicle and scapula. The only solid attachment of the arm–shoulder–pectoral girdle with the rest of the skeleton is at the sternoclavicular joint. The clavicle is a bony strut that holds the girdle, and thus the arm, away from the trunk. Movement occurs at the sternoclavicular joint and between the two members of the girdle at the acromioclavicular joint. Without these girdle joints permitting scapular and clavicular movements, the usefulness of the upper extremity would be greatly restricted. These two joints are strongly supported by ligaments. In general, when force is transmitted to the axial skeleton via a fall on the outstretched arm or against the point of the shoulder, fracture of the clavicle will occur rather than dislocation of either of the girdle joints.

Sternoclavicular joint. A strong joint, it rarely dislocates because of its tough strong anterior and posterior sternoclavicular ligaments, costoclavicular ligament and interclavicular ligament, and articular disc running from clavicle above to the first rib cartilage below. This is a mobile joint; circumduction is possible and it subtends all motion of the shoulder girdle.

Acromioclavicular joint. This allows movement in different directions between the two parts of the shoulder girdle. The joint itself is shallow (see x-rays and skeleton) and surrounded by a lax joint capsule ('acromioclavicular ligament'). The strength of the joint is in the tough and strong coracoclavicular ligament, which binds the clavicle to the scapula (base of coracoid) and thus keeps the acromioclavicular joint in place. Dislocation of this joint ('acromioclavicular separation') occurs but, as with the sternoclavicular joint, forces operating against the point of the shoulder more often result in fracture of the clavicle. Complete acromioclavicular separation (dislocation) requires tearing of both the coracoclavicular and acromioclavicular ligaments. Lesser degrees of separation also occur. These are serious injuries in terms of shoulder function. The key is the coracoclavicular ligament; as long as it remains intact, the shoulder can continue to function.

Muscles of the upper extremity.   These are divided into two groups: elevators (extensors) and supinators; and depressors (flexors) and pronators. All are supplied by spinal nerves C5–T1. These spinal nerves divide into anterior and posterior divisions in the formation of the brachial plexus. The posterior division supplies the elevators/ supinators; the anterior division supplies the depressors/pronators. This arrangement derives from the time when the upper extremity was a simple fin which, by virtue of its opposing muscles and separate nerves, could direct the creature up, down and around in the water.

Muscles of the scapula.   Muscles which hold the scapula in place and those that move the scapula include the rhomboids, levator scapulae, serratus anterior, trapezius and pectoralis minor.

Rhomboids
    Origin: Spinous processes of C6–7 and T1–4
    Insertion: Medial border of scapula
    Nerve: Dorsal scapular (from C5 root of brachial plexus)
    Action: Medial and upward movement of scapula

Levator scapulae
    Origin: Transverse processes of C1–4
    Insertion: Superior angle of scapula
    Nerve: C3–4 (cervical plexus, and branch of dorsal scapular nerve – C5)
    Action: Raises superior angle of scapula (rotates neck)

Serratus anterior
    Origin: Ribs 1–9
    Insertion: Vertebral margin of scapula
    Nerve: Long thoracic (C5–7 roots of brachial plexus)
    Action: Rotates scapula up and down, protracts scapula (elevates ribs)

Trapezius
    Origin: Occipital protuberance and neighboring occipital bone, ligamentum
        nuchae (C1–7), spinous processes of T1–12
    Insertion: Lateral third of clavicle, acromion, scapular spine
    Nerve: Spinal accessory
    Action: Rotates, adducts (draws it medially), raises or lowers the scapula (moves
        head to the side, rotates head to opposite side, extends the head)

Pectoralis minor
    Origin: Ribs 3–5
    Insertion: Coracoid process
    Nerve: Medial pectoral from brachial plexus (C8–T1)
    Action: Draws scapula forward and down (raises ribs in forced inspiration).

Trapezius is innervated by a cranial nerve. Trapezius was 'originally' a gill lifter, not a fin muscle. The gill muscles and branchial apparatus originated in branchial arches (*branchia* = Gk. gill); each arch required a nerve (arch 1, trigeminal; arch 2, facial; and so on). Branchogenic muscle trapezius (and its neighbor sternocleidomastoideus) is innervated by a branchial nerve, the spinal accessory.

Broad muscles are 'smart'; they may contract as a whole or in segments. Note the activity of the muscles acting on the scapula in Figure 6.1.

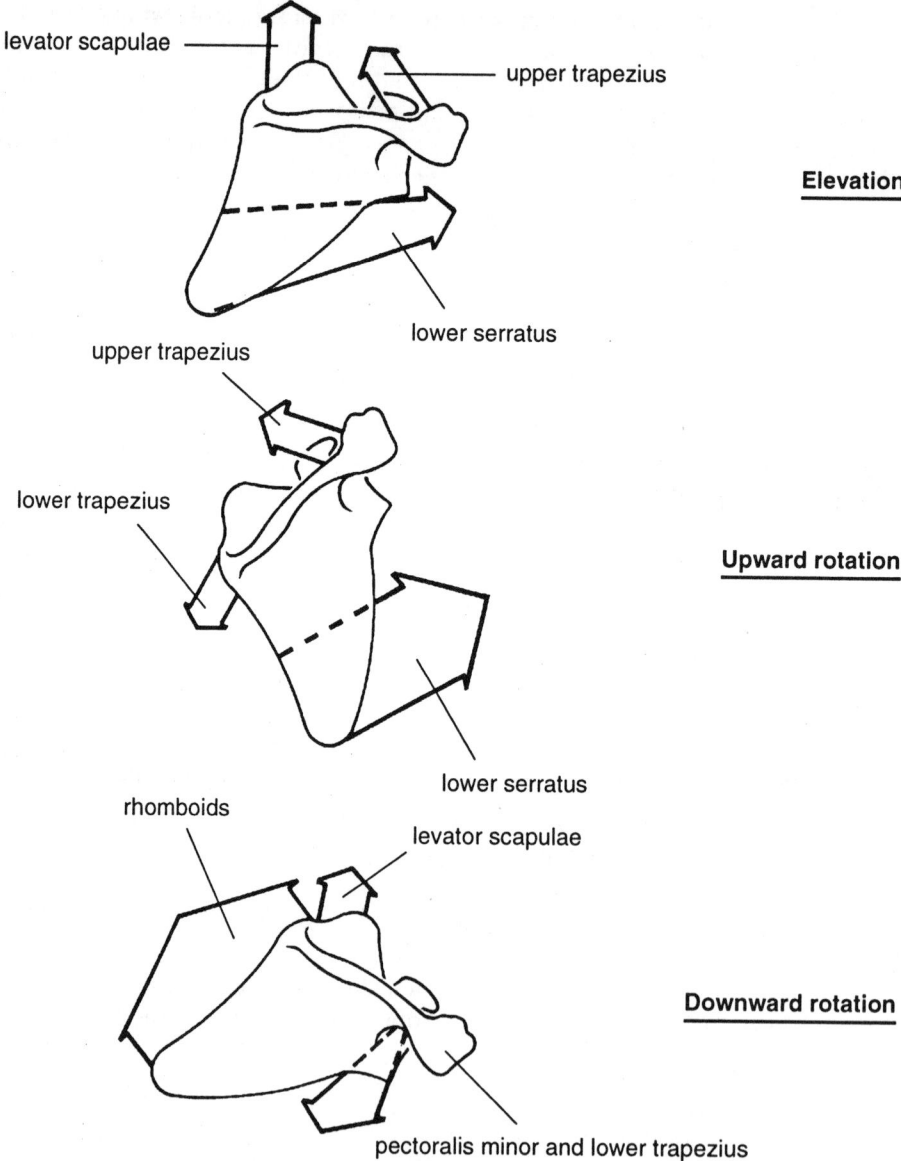

Figure 6.1 Activity of the muscles acting on the scapula

Glenohumeral joint. A ball-and-socket joint, the glenoid cavity of the scapula is shallow, but augmented by a substantial fibrocartilaginous rim, the **glenoid labrum**. The ligaments of import are:

Coracoacromial ligament. As it runs between two parts of the same bone, this ligament has no control on any joint motion. Rather, it forms a protective fibrous band over the top of the glenoid cavity by deepening the protective vault over the humeral head.

Joint capsule. The capsule is lax, but strong. The laxity permits a free range of motion. Its strength is augmented by several fibrous bands, the glenohumeral ligaments.

Tendon of long head of biceps. The tendon passes over the humeral head and is encased in a compartment of synovial membrane of the shoulder joint, in its course to the superior glenoid tubercle. By virtue of its close relation to the shoulder and the humerus itself, the tendon offers strong support.

Scapulohumeral muscles. The shoulder joint is strongly supported by these muscles which closely surround it (supra- and infraspinatus, teres minor, subscapularis and deltoid). The weakest area of the joint is the inferior aspect, where little muscular support is available. It is in this direction therefore that dislocation is most frequent, aided and abetted by the levering action of the humerus as it is abducted against the fulcrum of the acromion.

Muscles from the axial skeleton and pectoral girdle to the arm (humerus).

The chief action of these muscles is on the glenohumeral joint. Only major muscle activity is considered here. Group actions of the muscles operating the shoulder joint are complex and depend on various factors such as the initial position of the arm.

Activity produced:		
	Adduction	– Arm drawn in against side
	Abduction	– Arm elevated against ear
	Flexion	– Arm pointed straight forwards
	Extension	– Arm pointed straight backwards
	Medial rotation	– Arm turned medially
	Lateral rotation	– Arm turned outwards
	Circumduction	– Arm moved through a circle

Major adductors are pectoralis major, latissimus dorsi, teres major (actually part of latissimus dorsi) and subscapularis.

Pectoralis major
    Origin: Medial half of clavicle, sternum and costal cartilages of ribs 1-7
    Insertion: 5-cm tendon into crest of greater tubercle of humerus (anterior lip of bicipital groove)
    Nerve: Medial and lateral pectoral nerves from medial and lateral cords of the brachial plexus (C5–8 and T1)
    Action: Adduction and medial rotation of the arm, flexes arm

Latissimus dorsi
Origin: Spinous processes of T7–L5, sacrum and ilium (lumbodorsal fascia)
Insertion: Crest of lesser tubercle of humerus (posterior lip of bicipital groove)
Nerve: Thoracodorsal nerve from posterior cord of brachial plexus (C6–8)
Action: Adducts and medially rotates arm, extends arm

Teres major
Origin: Inferior angle of scapula
Insertion: Crest of lesser tubercle of humerus
Nerve: Branch of lower subscapular nerve from posterior cord of brachial plexus (C5–6)
Action: Same as latissimus dorsi (adducts and medially rotates arm, extends arm)

Subscapularis
Origin: Subscapular fossa of scapula
Insertion: Lesser tubercle of humerus
Nerve: Subscapular nerves (two) from posterior cord brachial plexus (C5–6)
Action: Adducts and medially rotates arm (the latter is main function).

Major abductors are supraspinatus and deltoid.

Supraspinatus
Origin: Supraspinous fossa of scapula
Insertion: Greater tubercle of humerus
Nerve: Branches of suprascapular nerve from superior trunk of brachial plexus (C5)
Action: Abduction of humerus

Deltoid
Origin: Lateral third of clavicle, acromion and scapular spine
Insertion: Deltoid, tubercle of humerus
Nerve: Axillary nerve, one of the terminal branches of posterior cord, brachial plexus (C5–6)
Action: Abducts arm; anterior portion flexes and medially rotates arm; posterior portion extends and laterally rotates arm.

Major flexors are the anterior portion of deltoid, pectoralis major, short head of biceps and coracobrachialis.

Short head of biceps
Origin: Coracoid process
Insertion: Tuberosity of radius
Nerve: Musculocutaneous, one of the terminals of lateral cord of brachial plexus (C5–6)
Action: Flexes arm (main activity is flexion of elbow and supination of forearm in company with long head of biceps)

Coracobrachialis
Origin: Coracoid process of scapula
Insertion: Medial aspect of body of humerus
Nerve: Musculocutaneous (C6–7)
Action: Flexes arm.

Major extensors are latissimus dorsi, teres major and posterior portion of deltoid.

Major medial rotators are teres major, latissimus dorsi, pectoralis major and subscapularis.

Major lateral rotators are infraspinatus and teres minor.

Infraspinatus
Origin: Infraspinous fossa of scapula
Insertion: Middle part of greater tuberosity of humerus
Nerve: Branches of suprascapular nerve of brachial plexus superior trunk (C5–6)
Action: Lateral rotation of humerus

Teres minor
Origin: Axillary border of scapula
Insertion: Lowest part of greater tubercle
Nerve: Branch of axillary nerve (C5)
Action: Lateral rotation of humerus.

As an example of the complex interaction of muscles to produce what appears to be a simple action, consider abduction of the arm. This first requires action of the rotator cuff muscles to hold the head of the humerus in the glenoid fossa. The head would otherwise be simply pulled up tight to the overlying acromion (and thereby fixed) as the abducting forces are applied. Then supraspinatus initiates abduction – the first 10 degrees or so – followed by deltoid, which continues the abduction to 90 degrees. At this point, the humerus impinges on the overlying acromion so that further abduction of the glenohumeral joint is impossible. To continue abduction and raise the arm alongside the ear, the entire scapula must be rotated through the sternoclavicular and acromioclavicular joints by action of superior trapezius fibers pulling on the lateral end of the scapular spine, lower trapezius fibers pulling down on the medial end of the spine, and serratus anterior pulling the rotated scapula up and around. No wonder the brachial plexus is so complicated! Not surprisingly, there is a considerable swelling in the cervical cord from C5 downwards to furnish neurons to all these muscle segments and, with the help of higher CNS centers, to organize a smooth interaction of all these structures.

An area of clinical importance, based on shoulder anatomy, involves the supraspinatus tendon. There is little room for this tendon to slide back and forth in between the acromion and greater tubercle of humerus to which it is attached. A bursa lies between supraspinatus and the acromion to reduce the friction. However, the supraspinatus tendon is frequently inflamed and torn by the constant squeezing through this tight area. The inflamed tendon then inflames the overlying bursa. Everyone is familiar with the painful problem of subacromial bursitis.

141

Elbow joint. A hinge joint; the rounded trochlea of the humerus articulates with the semilunar notch of the ulna, and the capitulum of the humerus articulates with the fovea atop the head of the radius. The coronoid process of the ulna passes into the coronoid fossa of the humerus in flexion; the olecranon process of the ulna passes into the humeral olecranon fossa in extension. Strong radial and ulnar collateral ligaments extending from the humeral epicondyles to the radius and ulna support the joint.

Radioulnar joints. These pronate and supinate the wrist/hand; two joints are necessary to allow the radius to roll over the ulna.

Proximal radioulnar joint. This is enclosed within the articular capsule synovia of the elbow joint. The round disc-like head of radius is held against the neighboring radial notch of the ulna by an **annular ligament** which is synovial-lined, the whole permitting free spinning of the radius head. The radius head is in close proximity to the important deep branch of the radial nerve. Surgeons, beware!

Distal radioulnar joint. The second pivot joint between the two bones, this lies between the head of ulna and ulnar notch of the distal radius. Dorsal and volar radioulnar ligaments are present. A good-sized articular disc extends from the radius to the ulnar styloid, binding the two bones together. The proximal surface of the disc is jointed with the ulna; the distal surface is jointed with the carpal triquetrum and medial edge of lunate. The whole is enclosed in a synovial-lined joint capsule.

Muscles of the arm. There are only three and they operate the elbow.
Posterior extensor (elevator): Triceps (three-headed)
Anterior flexors (depressors): Biceps (two-headed) and brachialis

Triceps
Origin: Long head: Infraglenoid tubercle of scapula
Lateral head: Posterolateral body of upper humerus
Medial head: Posterior body of lower humerus
Insertion: Olecranon of ulna
Nerve: Radial nerve of brachial plexus posterior cord (C6–7)
Action: Extends the forearm

Biceps
Origin: Long head: Supraglenoid tuberosity [(tendon runs through the glenohumeral joint and rests in the intertubercular groove of upper humerus (bicipital groove)]
Short head: Apex of coracoid process
Insertion: Radial tuberosity of medial upper radius (plus a broad aponeurosis extending medially over the brachial artery to the deep fascia of the forearm flexors)
Nerve: Musculocutaneous of lateral brachial plexus cord (C5–6)
Action: Flexes the forearm; supinates the hand

Brachialis
Origin: Lower half of front of humerus
Insertion: Tuberosity of ulna
Nerve: Musculocutaneous (C5–6)
Action: Flexes the forearm.

The biceps belongs to the depressor (flexor)–pronator muscle group but, due to its medial radius insertion, it proves to be a strong twister (supinator) of the forearm, an exception to the general rule. The forearm flexors are innervated by the lateral cord of the brachial plexus – the musculocutaneous nerve. Damage to the lateral cord or musculocutaneous nerve will not completely eliminate elbow flexion as the brachioradialis muscle is also capable of this motion, another exception to the general rule, as the brachioradialis belongs to the elevator (extensor) muscle group.

Tears of roots C5–6 may occur in difficult deliveries (shoulder presentations) or in falls where the head and neck go one way and the shoulder another (for example, a motorcyclist 'diving' into a tree). With loss of C5–6, the position of the upper limb depends on the unopposed pull of muscles that are normally opposed by muscles innervated by C5–6 (the proximal upper extremity muscles – the rotator cuff muscles, deltoid, biceps, brachialis and coracobrachialis). With loss of the external rotators, the arm is internally rotated by pectoralis major and latissimus dorsi, and hangs limply by the side due to loss of supraspinatus and deltoid abductors. Loss of biceps results in pronation of the hand, called the 'waiter's tip' position. The paralysis is more elegantly called Erb-Duchenne paralysis.

Another birth injury of the brachial plexus roots that may occur in difficult deliveries involves hyperabduction of the arm. The same type of root injury occurs when falling from a tree and grabbing a passing limb. The roots torn are the lower ones – C8–T1. Loss of these roots is most manifest in the medial cord of the plexus and its branches. The posterior cord radial nerve also contains C8–T1, but admixture of C5–7 in the radial nerve wards off significant loss of extensor function. As C8–T1 form the ulnar nerve, and these roots provide the fibers in the median nerve running to the hand, varying degrees of palmar and digital numbness and palsy of the intrinsic muscles of the hand will be found, depending on the extent of damage to these C8–T1 roots. Metacarpophalangeal joints are hyperextended and interphalangeal joints hyperflexed. Considerable wasting in the hand becomes obvious. The condition is termed Klumpke's paralysis.

Axilla. The axilla is a pyramidal space. The base of the pyramid is the axillary skin (armpit); anteriorly are the pectoral muscles, posteriorly are latissimus dorsi, teres major and subscapularis; medially are serratus and the chest wall, and laterally is the bicipital groove of the humerus containing the tendon of the long head of biceps. The anterior and posterior muscular axillary walls insert on the anterior and posterior lips (crests of the greater and lesser humeral tuberosities) of the bicipital groove. The apex of the axilla lies at the crossing of the clavicle and first rib. The axilla is the important container of the blood, lymphatic and nerve supplies of the entire upper extremity.

143

Axillary artery. This runs from the border of the first rib (entry into axilla) to the lateral margin of teres major (lateral margin of axilla). It may be divided into three parts by the crossing of the pectoralis minor muscle over its midportion.

Branches are:

First part: The small, highest thoracic artery, which supplies blood to the first and second intercostal spaces. (The first intercostal artery from the aorta runs to the third intercostal space).

Second part: **Thoracoacromial artery**, which supplies blood to the pectoral muscles, and muscles and tissues around the acromion; and **lateral thoracic artery,** which descends along the lateral aspect of pectoralis minor supplying blood to this muscle and serratus anterior;

Third part: The **large subscapular artery**, which proceeds down along the axillary border of scapula and subscapular muscle. Its two main branches are:

Circumflex scapular, which winds around the scapula to meet and anastamose with branches of the suprascapular artery from the subclavian (see Neck section);

Thoracodorsal artery, which is in fact the continuation of the subscapular below the take-off of the circumflex scapular, and supplies subscapularis, latissimus dorsi and lower serratus anterior;

Anterior and posterior circumflex humeral arteries, which encircle the neck of the humerus and supply the humeral head, glenohumeral joint and muscle covering the humerus in this area.

Suprascapular artery. A branch of the subclavian thyrocervical trunk, runs transversely along the superior border of the scapula, then dives downwards to supply the posterior aspect of the scapula and its supra- and infraspinatus muscles. If the axillary artery is occluded, blood may be delivered to the arm via the connection between the suprascapular and the circumflex scapular, with reversal of the usual direction of flow in the latter vessel.

Axillary vein. Runs alongside the artery. A tributary, the **cephalic** vein, lies in the groove between deltoid and pectoralis major. At this location, this vein may easily be dissected and used as a venous access route.

Axillary lymphatics. Lymph from the entire upper extremity, chest wall, breast and lower neck all drains into lymph nodes clustered in the axillary fat with trunks that accompany the axillary vessels. This is a major lymph depot! Efferents from the axilla run with the axillary vein to empty into the junction of the jugular and subclavian veins.

Nerves of the axilla: The brachial plexus (Figure 6.2). The nicely symmetrical C5–8 and T1 spinal segmental nerves undergo convolutions, twists and turns to keep up with the twists and turns of the structures of the upper extremity as they developed from a simple fin to the complex human limb. The result is a brachial plexus of nerves clustered around the axillary artery. Every physician must know the anatomical make-up of the

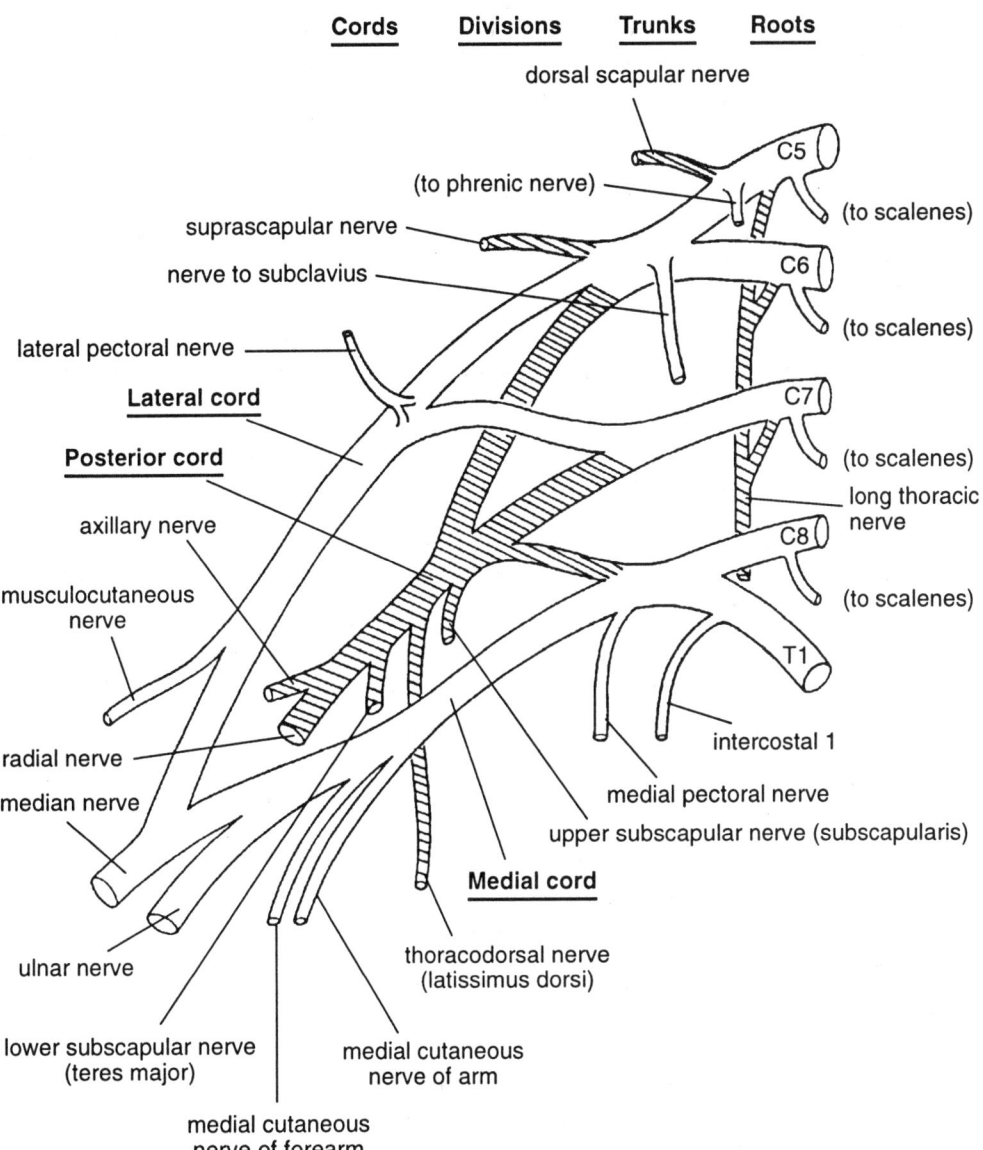

Figure 6.2   Brachial plexus

brachial plexus and its branches to understand the neurological problems that so often affect the function of the upper limb.

The brachial plexus should be memorized – roots, trunks, divisions, cords and branches. These are all ventral primary rami of the C5–8 and T1 spinal nerves. (Dorsal primary rami supply dorsal back musculature and skin. Longus colli and the scalenes are ventral muscles.) Roots, trunks and divisions of the plexus are formed at supraclavicular levels. The axilla contains the cords and their branches. Note the **long thoracic nerve** from C5– 7. This important nerve runs along the medial wall of the axilla, passing behind the other

145

neurovascular structures of the axilla, to innervate serratus anterior. This important muscle was in use when upper limbs were fins, before the development of a complex nerve plexus to satisfy the demands of the more complex limb. It should be no surprise that this nerve originates from cervical nerve roots proximal to the brachial plexus proper. Injury to the long thoracic nerve results in paralysis of serratus anterior, with accompanying difficulty in full abduction and protraction plus the cosmetically undesirable 'winging' of the scapula.

Other important nerve branches to the pectoral girdle muscles given off in the supraclavicular area above the axilla are the **dorsal scapular** from root C5 innervating the rhomboids, and the **suprascapular nerve** from the superior trunk (C5–6) innervating supra- and infraspinatus muscles. The course of the latter nerve is along the superior border of the scapula in company with the supraclavicular vessels to a notch in the superior scapula. The nerve passes through the notch to reach the posterior scapular muscles. Again, be aware that these proximal branches run to muscles operating the 'base of the fin' and are not part of the plexus distributing nerves to the later-arriving muscles of a complex upper extremity.

Separation into anterior and posterior divisions is required to furnish nerves to anterior and posterior division muscles (anterior depressors/flexors/pronators and posterior elevators/extensors/supinators). These muscles are farther down the evolutionary trail; their nerves are formed farther out in the brachial plexus. The pectoral muscles are depressors – anterior division nerves. Latissimus dorsi, teres major and subscapularis are elevators – posterior division nerves. Farther down the extremity, the depressor/flexor and elevator/extensor nature of the musculature is obvious.

Cord designations indicate the relationship to the axillary artery: medial cord on medial aspect of the artery (against the chest wall); lateral cord on the lateral side of the artery; and posterior cord on the back side of the artery.

Knowing the root origins of the cords allows you to at least approximate the root levels of the nerves supplying the upper extremity muscles. The **lateral cord** is formed from anterior divisions of C5–7; the **lateral** pectoral nerve contains fibers from these roots. The **medial cord** is from C8–T1; the **medial** pectoral nerve contains C8–T1 fibers. All roots are represented in the posterior cord. 'Higher' roots go to proximal muscles. Thus, C5–6 root fibers do most of the work for the scapular and proximal arm muscles. (Supra- and infraspinatus muscles are innervated by a nerve given off proximal to the trunk divisions, the supraclavicular nerve. This, too, is C5–6.)

Nerves and vessels receive a covering of prevertebral deep fascia as they cross the lower neck and pass into the axilla. The fascia forms a definite axillary sheath which binds them all together. Local anesthetic block of the brachial plexus may be obtained by injecting the anesthetic agent into the sheath, where it is confined, and bathes and anesthetizes the plexus nerves.

There are two terminal branches of each cord: musculocutaneous and lateral root of the median nerve from the lateral cord C5–7; ulnar and medial root of the median nerve from the medial cord C8–T1; and the axillary and radial nerves from the posterior cord C5–8 and T1. The median nerve itself contains C6–8 and T1. The position of the terminal branch nerves – axillary, radial, musculocutaneous, median and ulnar – throughout their upper extremity course should be known by every clinician in their three-dimensional relationships. They are described here for continuity, although this section is still 'in the axilla'. Commit these anatomical points to memory!

Radial nerve. This continues posterior to the axillary artery to the border of teres major, then spirals around the humerus from medial to lateral side of the bone to emerge at the elbow, passing anterior to the lateral epicondyle of the humerus under cover of the brachioradialis muscle. Progressing peripherally under cover of this muscle, the radial nerve divides into deep and superfical branches at the level of the head of radius. The deep branch penetrates the supinator muscle and passes around the lateral radius to reach the dorsal aspect of the interosseous membrane as the dorsal interosseous nerve. The superficial branch continues down the forearm beneath brachioradialis. The radial nerve supplies the elbow, wrist and digit extensors, supinator, thumb elevators (extensors and abductor longus) and brachioradialis. (The latter flexes the elbow joint, an exception to the rule of extensor function for posterior division muscles.) Skin branches of the radial nerve supply the posterior aspects of the arm, forearm and hand.

Axillary nerve. This branches from the posterior cord to pass in front of the subscapularis tendon, and between teres minor and major, to reach the posterior aspect of the neck of humerus through a quadrilateral space bordered by teres minor and major, long head of triceps and humeral neck. The axillary nerve supplies teres minor, deltoid and the overlying skin.

Musculocutaneous nerve. This branches from the lateral cord at the border of pectoralis minor, pierces coracobrachialis and comes to lie between biceps and brachialis during its course in the arm. Just above the elbow, the musculocutaneous nerve leaves its comforable intermuscular position to become the lateral antibrachial cutaneous nerve. It innervates the three flexors of the arm.

Median nerve. The 'nerve of the forearm and thumb', the median does nothing in the arm. It is formed from median branches of the lateral and medial cords, and lies ventral to the artery in the axilla. The median nerve passes down the arm in company with the brachial artery covered only by the deep brachial fascia. At the elbow, it lies medial to the artery and over the medial humeral epicondyle. Piercing pronator teres in the proximal forearm, the nerve passes peripherally between deep and superficial finger flexors to the wrist. In the forearm, it gives off the volar interosseous nerve, which courses along the volar aspect of the interosseous membrane. At the wrist, the nerve lies directly under the palmaris longus tendon, and between the superficial flexor and flexor carpi radialis tendons. After passing through a tunnel beneath the transverse carpal ligament, it almost immediately emits the important recurrent muscular branch to the thenar eminence muscles. The nerve then breaks up into branches to the thumb, index and middle fingers, and lateral half of the ring finger. Muscles innervated by the median nerve include all forearm-based flexors of the wrist and fingers except flexor carpi ulnaris and the medial half of flexor digitorum profundus, which are handled by the ulnar nerve. In the hand, the median nerve innervates the thenar eminence muscles and lumbricals of the index and middle fingers.

Ulnar nerve. The 'nerve of the hand', this begins with the median nerve and brachial artery in the upper arm, then gradually sinks back to pass across the elbow

147

in the familar 'funny bone' position posterior to the medial epicondyle of humerus. Under cover of the flexor carpi ulnaris, the ulnar nerve proceeds down the forearm to cross the wrist between the pisiform and hook of hamate. In the hand, it supplies all the muscles of the hypothenar eminence, and all the intrinsic muscles of the hand except the thenar muscles and the first two lumbricals. The ulnar nerve also supplies adductor pollicis and deep head of flexor brevis, which is in fact an interosseous muscle. In the forearm, the ulnar innervates flexor carpi ulnaris and the ulnar half of flexor digitorum profundus. Skin representation of the ulnar comprises the little finger and medial half of the ring finger, and the corresponding medial palm and dorsum of the hand.

Arteries of the arm.
The **brachial artery** is the continuation of the axillary artery starting at the border of teres major and continuing to the proximal forearm, where it divides into radial and ulnar arteries. Its course is down the medial aspect of the arm covered only by brachial fascia and skin. The brachial pulse is palpable throughout its extent. The artery is initially accompanied by the ulnar, median and radial nerves, but the radial soon drops down to wind around the humerus and the ulnar drops back to the medial posterior area, leaving the median nerve as the only true 'friend' of the artery all the way down the arm. Brachial artery branches of note are the profunda, and the superior and inferior ulnar collaterals. (Radial collaterals stem from the profunda.)

Profunda brachii. This branches off high in the arm and accompanies the radial nerve in its spiral course around the humerus. Just above the elbow, the artery divides into the **radial collateral** and **middle collateral arteries.** These collaterals meet radial recurrent and ulnar interosseous recurrent arteries on the lateral side of the elbow.

Superior and inferior collaterals. These brachial artery branches arise just above the elbow (the superior collateral accompanies the ulnar nerve) and anastomose with anterior and posterior recurrent vessels from the ulnar artery on the medial side of the elbow.

The collateral circulation around the elbow mirrors the collateral vascular circulation around the knee. Vascular problems, however, are much more serious and common in the heavily muscled weight-bearing lower extremity than in the elbow and upper extremity.

Muscles of the forearm.
Flexors (pronators) arise from the medial epicondyle of humerus (and, lower down, anterior surfaces of radius, ulna and interosseous membrane) and are supplied by anterior division brachial plexus nerves; extensors (supinators) arise from the lateral epicondyle of humerus (posterior surfaces of radius, ulna and interosseous membrane) and are supplied by posterior division plexus nerves. Flexors and extensors each have superficial and deep groups.

Superficial flexors are pronator teres, flexor carpi radialis, palmaris longus, flexor carpi ulnaris and flexor digitorum superficialis.

Pronator teres
  Origin: Medial epicondyle of humerus and coronoid process of ulna (two heads;
    median nerve passes between the two heads)
  Insertion: Mid-radius
  Nerve: Median (C6–7)
  Action: Pronates hand

Flexor carpi radialis
  Origin: Medial epicondyle of humerus
  Insertion: Through a separate compartment of the flexor retinaculum to the base
    of the third metacarpal
  Nerve: Median (C6–7)
  Action: Flexes and abducts hand

Palmaris longus (unimportant and inconstant muscle useful for tendon grafts)
  Origin: Medial epicondyle of humerus
  Insertion: Palmar aponeurosis
  Nerve: Median (C6–7)
  Action: Tenses palmar aponeurosis, flexes hand

Flexor carpi ulnaris
  Origin: Medial epicondyle of humerus and olecranon
  Insertion: Pisiform
  Nerve: Ulnar (C8–T1)
  Action: Flexes and adducts hand

Flexor digitorum superficialis
  Origin: Medial epicondyle, coronoid process of ulna, medial aspect of upper radius
  Insertion: Its four tendons pass through the carpal tunnel, cross the palm and
    insert on the sides of the middle phalanges. At the base of the proximal
    phalanges, each tendon splits to allow passage of the deep flexor tendons. After
    briefly reuniting to form a channel for the accompanying flexor profundus
    tendon, each tendon splits again to insert on both sides of the middle phalanges
  Nerve: Median (C7–8 and T1)
  Action: Flexes middle phalanx; by continued action, flexes first phalanx and then
    the hand itself.

Deep flexors are flexor digitorum profundus, flexor pollicis longus and pronator
quadratus.

Flexor digitorum profundus
  Origin: Superior half of ulna and interosseous membrane
  Insertion: Four tendons pass through the carpal tunnel, across the palm, and
    through the split superficialis tendons at the level of the base of the first
    phalanges to insert in the bases of the terminal phalanges
  Nerve: Lateral half by the volar interosseous of median (C8–T1) and medial half
    by the ulnar (C8–T1)
  Action: Flexes distal phalanx; by continued action flexes other phalanges and hand

149

Flexor pollicis longus
    Origin: Upper body of radius and interosseous membrane
    Insertion: Distal phalanx of thumb
    Nerve: Volar interosseous of median (C8–T1)
    Action: Flexes second phalanx of the thumb; by continued action flexes first phalanx
        and first metacarpal

Pronator quadratus
    Origin: Distal ulna, anterior surface
    Insertion: Distal radius, anterior surface
    Nerve: Volar interosseous of median (C8–T1)
    Action: Pronates hand.

Superficial extensors are brachioradialis, extensor carpi radialis longus and brevis,
extensor digitorum communis (and extensor digiti quinti) and extensor carpi ulnaris.

Brachioradialis
    Origin: Border of humerus above lateral epicondyle
    Insertion: Styloid process of radius
    Nerve: Radial (C5–6)
    Action: Flexes forearm (Brachioradialis was once a long supinator doing what
        posterior muscles are supposed to do. With the development and twisting of
        the forearm, it lost its supination power and became a 'bastardized' flexor while
        retaining its posterior division nerve, the radial)

Extensor carpi radialis longus and brevis
    Origin: Lateral epicondyle of humerus
    Insertion: Longus to second metacarpal; brevis to third metacarpal
    Nerve: Radial (C6–7)
    Action: Extend and abduct hand

Extensor digitorum communis (and digiti quinti)
    Origin: Lateral epicondyle of humerus
    Insertion: Proximal phalanges and, by tendinous slips, the middle and distal
        phalanges (see Hand section)
    Nerve: Deep radial (C6–8)
    Action: Extend digits (see Hand section)

Extensor carpi ulnaris
    Origin: Lateral epicondyle and posterior surface of ulna
    Insertion: Base of fifth metacarpal
    Nerve: Deep radial (C6–8)
    Action: Extends and adducts hand.

Deep extensors are supinator, abductor pollicis longus, extensor pollicis brevis, extensor
pollicis longus and extensor indicis proprius.

Supinator
> Origin: Lateral epicondyle, radial collateral ligament and supinator crest of proximal ulna
> Insertion: Medial surface of radial tuberosity and down lateral border of proximal radius
> Nerve: Deep radial (C6)
> Action: Supinates hand

Abductor pollicis longus
> Origin: Interosseous membrane and adjoining surfaces of radius and ulna
> Insertion: Base of first metacarpal
> Nerve: Deep radial (C6–7)
> Action: Abducts thumb at carpometacarpal joint

Extensor pollicis brevis
> Origin: Interosseous membrane and radius below origin of abductor longus
> Insertion: Base of proximal phalanx of thumb
> Nerve: Deep radial (C6–7)
> Action: Extends proximal phalanx

Extensor pollicis longus
> Origin: Ulna below origin of abductor longus and from the interosseous membrane
> Insertion: Base of terminal phalanx of thumb
> Nerve: Deep radial (C6–8)
> Action: Extends terminal phalanx of thumb

Extensor indicis proprius
> Origin: Ulna below extensor pollicis longus and from the interosseous membrane
> Insertion: Joins common extensor of index finger at level of second metacarpal head
> Nerve: Deep radial (C6–8)
> Action: Extends index finger.

Just as the great toe is the important digit of the foot, the thumb is THE important digit of the hand. Both great toe and thumb have special muscles for their use. The all-important thumb has a long abductor and short extensor in addition to the long extensor and long flexor. The three 'outcropping' muscles – abductor longus, and extensors longus and brevis – are all deep muscles innervated by the deep branch of the radial. Remember, the deep branch passes down to these muscles by slipping through the supinator muscle. Supinator may not be essential (biceps is also able to supinate), but the deep radial nerve is. Without it, function of these outcropping muscles is lost and **the hand loses much of its human function.**

These three important outcropping muscles of the thumb outline a space referred to as the 'anatomical snuffbox' at the base of the thumb. The dorsal tendon of the box is the extensor pollicis longus; the volar border tendons are abductor pollicis longus and extensor pollicis brevis. The radial artery runs across the floor of the snuffbox.

151

As in movements of the shoulder, synergistic multiple muscle activity is necessary to produce the desired movements in the hand. For instance, when you clench your fist tightly, note the activity of the flexor tendons AND extensor muscles of the forearm. To produce a decent fist, primarily a flexor activity, extensor muscles must tighten concomitantly to counteract the flexion of the wrist which would otherwise occur as a continued flexor action after flexion of the digits. The reverse synergistic action occurs with extensor pull on the proximal phalanges.

Muscle groups in the forearm are separated into compartments by tough intermuscular septa just as in the muscles of the leg. The interosseous septum separates dorsal and volar extensor and flexor groups. Septa are present between superficial and deep groups; and the binding external muscle fascia is also tough. Thus, these are non-distensible compartments in the forearm (as in the leg). Compartment swelling caused by bleeding, inflammation, etc., may require relief by fascial incision to protect the viability of the contained tissue.

Arteries of the forearm.   The collateral circulation around the elbow has already been described. The brachial artery divides into its radial and ulnar branches approximately 1 cm into the forearm. The ulnar is the larger, and busier, artery in the forearm.

Ulnar artery.   This courses down the forearm under cover of flexor carpi ulnaris in company with the ulnar nerve. It crosses the transverse carpal ligament just inside the pisiform bone, and immediately divides into **deep** and **superficial branches** of the deep and superficial palmar arches (see Radial artery section below). In its course through the forearm distal to its recurrent branches around the elbow, the **common interosseous** artery is given off just below the level of the radial tuberosity. This vessel passes deeply for around 1 cm and, when it reaches the upper border of the interosseous membrane, divides into **anterior** and **posterior interosseous arteries** (*cf* similar branching in the lower extremity). These vessels supply the deep muscles on either side of the membrane. The median nerve accompanies the anterior interosseous, and the deep radial nerve posterior interosseous branch accompanies the posterior interosseous artery. The anterior interosseous emits a recurrent branch which anastomoses with the middle collateral branch of the profunda brachii.

Radial artery.   Immediately after its formation by the bifurcation of the brachial, the radial artery emits the radial recurrent artery, which runs proximally to anastomose with the radial collateral branch of the profunda brachii. The radial artery passes to the wrist under cover of the brachioradialis muscle. It then winds backwards around the lateral carpus, passing across the floor of the snuffbox, to reach the space between the thumb and index metacarpals. It then turns medially, passes deep in the palm across the bodies of the metacarpal bones, and joins the deep branch of the ulnar to form the **deep palmar arch**. The radial constitutes the main component of the deep arch. Before entering the snuffbox, a superficial palmar branch is given off which anastomoses with the deep palmar branch of the ulna. This completes the **superficial palmar arch** lying between the palmar aponeurosis and the superficial flexor tendons. The ulnar artery is the main contributor to the superficial arch. If the arches are complete, ligation or occlusion of either the radial or ulnar will not permanently damage the circulation of the hand.

Carpal tunnel. The median nerve and all flexor tendons (except for flexor carpi ulnaris) reach the hand by passing through the carpal tunnel, which is formed by the flexor retinaculum stretching from the pisiform bone and hook of the hamate to the tuberosities of scaphoid and trapezium. No major arteries pass through the tunnel. The tendons in the tunnel are encased within synovial sheaths for easy gliding. The familiar 'carpal tunnel syndrome' is caused by the swelling of inflamed tendons exerting pressure on the median nerve in this tight indistensible space.

Hand. Intricate nerve, muscle, tendon and joint arrangements in the hand make possible the dexterous activity performed only by humans, activities that are essential for normal human unskilled and skilled life. Awareness of the capabilities and necessary performance of this marvelous instrument instills all physicians with a respect and sense of responsibility for its health. Only by understanding the basic anatomy of the hand can a doctor competently assume this responsibility.

In general, the normal healthy hand possesses tough palmar skin akin to the skin of the sole of the foot. Digital skin is especially well-endowed with sensory nerve endings, making the terminal phalangeal skin pads exquisitely sensitive. The thumb is rotated to lie at a right angle to the plane of the fingers, thus providing a pinch mechanism. Dextrous and/or strong and/or rapid digital activity is possible by the presence of intrinsic hand muscles as well as the long-tendoned forearm muscles. Muscle motor units are small, requiring a rich supply of nerve fibers. Note that the thumb is at least half the hand. Without the thumb, no pinch action is possible.

Flexor tendon sheaths. Flexor tendons must twist around many corners as joints flex and extend; they must glide effortlessly and painlessly as they shorten and lengthen; they must be held in place to prevent 'bowstringing' when contracted. To make all this possible, the flexor tendons travel over the phalangeal joints surrounded by slippery synovial sheaths which in turn are encased in sheaths of fibrous tissue (*cf* pericardium). The whole is held down firmly by fibrous bands attached to the bony sides of the phalanges. The sheaths for digits 2–4 extend from the terminal phalanx to the distal palm just proximal to the metacarpophalangeal joints. The sheaths for the thumb and small finger extend to the distal forearm. At the level of the flexor retinaculum, the latter expand to envelop tendons 2–4 so that all the flexors are again ensheathed as they pass through the carpal tunnel over the many carpal joints.

Lacerations of digital sheaths or contained tendons, or infection within the sheaths may wreck the gliding function of the tendons because of scarring, adhesions and narrowing. A stiff immovable finger is relatively useless! Prompt care of the infected tendon sheath can minimize the consequent difficulties. Often, injured tendons must be replaced by transplants (for example, plantaris or palmaris longus tendons) to restore function.

Extensor tendon sheaths.   These tendons are also encased in synovial sheaths as they pass under the dorsal wrist extensor retinaculum, a fibrous wristlet that holds these tendons in place.

Insertion of forearm tendons.

Flexors.   The arrangement of superficial and deep flexor tendons in the digits has already been described. To review, the superficial flexor tendons divide at the base of the first phalanx into two slips to allow passage of the deep flexor tendon, which must attain the terminal phalanx. Once the deep tendon is through, the divided superficial tendons reunite and form a floor for the now overlying profundus tendon. The superficialis tendon finally splits again and inserts into the sides of the middle phalanx.

Extensors.   This is a complicated affair! The central part of the extensor communis tendon inserts in the base of the proximal phalanx, making possible the chief function of this muscle, extension of the metacarpophalangeal joint. However, around this insertion and covering the dorsolateral aspects of the proximal phalanx, the extensor tendon expands into a fibrous aponeurotic 'hood'. At its apex just proximal to the proximal interphalangeal joint, the hood splits into three parts: a central slip running to the base of the second phalanx; and two lateral slips running to the sides of the terminal phalanx. The hood is pulled back over the metacarpophalangeal joint during extension, but is pulled distally over the proximal phalanx during flexion. Interestingly, the forearm digital extensor tendon exerts almost its entire pull on the metacarpophalangeal joint but, after it has extended the joint, it has little (shortening) power left to extend the interphalangeal joints via the central and lateral slips of the extensor expansion. Thus, this task is performed by the intrinsic lumbricals and interossei (see below).

Intrinsic muscles of the hand.

Thenar muscles:	Abductor pollicis brevis
	Flexor pollicis brevis
	Opponens pollicis
	Adductor pollicis
Hypothenar muscles:	Abductor digiti quinti
	Flexor digiti quinti brevis
	Opponens digiti quinti
Lumbricals:	Four
Interossei:	Palmar interossei (three)
	Dorsal interossei (four).

A total of 18 muscles, all are innervated by the ulnar nerve except the three thenar eminence muscles and the lumbricals to the index and middle fingers, which are served by the median nerve.

Thenar muscles. All arise from the transverse carpal ligament (flexor retinaculum).
Abductor pollicis brevis
Origin: Flexor retinaculum
Insertion: Radial side of proximal phalanx of thumb
Nerve: Recurrent branch of median (C8–T1)
Action: Abducts thumb

Flexor pollicis brevis
Origin: Flexor retinaculum
Insertion: Radial side of proximal phalanx
Nerve: Recurrent branch of median (C8–T1)
Action: Flexes and abducts thumb (deep portion of flexor pollicis brevis is in fact an interosseous muscle and has ulnar innervation)

Opponens pollicis
Origin: Flexor retinaculum
Insertion: Radial border of thumb metacarpal
Nerve: Recurrent branch of median (C8–T1)
Action: Abducts, flexes and rotates (opposes) metacarpal of thumb; brings the thumb out in front of the palm to face the fingers.

The above three muscles are the 'muscles of the thenar eminence'.

Adductor pollicis
Origin: Third metacarpal
Insertion: Ulnar side of base of first thumb phalanx
Nerve: Ulnar (C8–T1)
Action: Adducts thumb.
This muscle lies deep in the palm, deep to the flexor tendons and lumbricals.

Thumb movements.
Flexion: Bending the thumb across the surface of the palm to the pad over the fifth metacarpophalangeal joint
Extension: Drawing the thumb away from the palm in the plane of the palm
Abduction: Drawing the thumb away from the palm at a right angle to the palm
Adduction: Drawing the thumb back against the palm from an abducted position
Opposition: Flexing, adducting and rotating the thumb so that its pad directly opposes the pad of any of the other digits
Circumduction: Combination of all of the above so that the thumb (metacarpal) moves in a circle.

Lumbricals (L. worm).   Four worm-like muscles originating from the four flexor profundus tendons in the proximal palm, and passing distally to insert in the radial sides of the extensor hoods and lateral bands of the same digits empowered by their parent tendons. Thus, they are ultimately attached to the middle and terminal phalanges. Lumbricals to the index and middle fingers are innervated by the median nerve (C8–T1) whereas the two to the ring and little fingers are innervated by the ulnar nerve (C8–T1). (The action of the lumbricals is described below after description of the interossei muscles.)

Interossei.   Innervation to all interossei is by the ulnar nerve (C8–T1). Palmar and dorsal interossei have insertions both into the proximal phalanges and into the extensor hood and lateral slips.

Palmar interossei.   Three in number, they originate from the volar surfaces of the metacarpals of the index, ring and little fingers. Bony insertion is into the bases of the proximal phalanges of index, ring and little fingers. The index phalanx bony insertion is on the ulnar side; ring and little digit bony insertion is on the radial side. Thus, with the fingers in extension, palmar interossei contraction adducts the digits toward the middle finger.

The second insertion of these interossei is into the extensor expansion hood and its lateral slips. Details of this action, which is in concert with the hood slip insertions of the dorsal interossei and lumbricals, is detailed below.

Dorsal interossei.   Four altogether, their origin is from adjacent sides of the metacarpals (thumb-index, index-middle, middle-ring and ring-little). These then are bipennate (two-feathered) muscles. Insertion is as with the palmar group: a bony insertion into the proximal phalangeal bases, and an insertion into the extensor hood and lateral slips. Bony insertions of the thumb-index and index-middle muscles are into the radial side of the index and middle proximal phalanges. Those of the middle-ring and ring-little are into the ulnar sides of the middle- and ring-finger proximal phalanges. Thus, with the fingers extended, dorsal interossei fan or abduct the fingers. (The middle digit can be moved both ways: towards radial or ulnar. Abductor digiti quinti and the thumb abductors 'fan' the two 'outside' digits.)

The second type of insertion involves the lumbricals and all of the interossei. Note that the extensor hood over the proximal phalanges moves back over the metacarpophalangeal joint towards the wrist with extensor communis muscle contraction. Conversely, it is pulled distally away from the metacarpophalangeal joint with extensor relaxation and finger flexion. The extensor apparatus has farther to go in flexion as it lies on the outside of the flexed finger curve. With the metacarpophalangeal joints extended and the hood thus pulled centrally, pull on the hood by the intrinsics extends the interphalangeal joints through pull on the central and lateral hood slips. (Extensor communis alone cannot put much pull tension on the interphalangeal joints. Its main function is extension of the metacarpophalangeal joints.) On the other hand, with the hood pulled distally by finger flexion and the required relaxation of antagonistic extensor pull, pull of the intrinsics through their attachments to the hoods and proximal phalanges is then at 90° to the axis of the proximal phalanges. Strong flexion of the metacarpophalangeal joints results. To summarize, the intrinsics extend the interphalangeal joints in extension mode and flex the metacarpophalangeal joints in flexion mode.

Digital blood and nerve supply.   The palmar vascular arches have already been described. The superficial palmar arch, mostly ulnar artery, is covered only by skin and the palmar aponeurosis. It lies beneath an imaginary line crossing the palm at the level of the distal border of the fully extended thumb. The deep arch, mostly radial artery, lies deep in the proximal palm, running across the proximal metacarpals and palmar interossei deep to the flexor tendons. The deep arch is approximately 1 cm proximal to the level of the superficial arch. Digital vessels are given off from the arches.

Digital nerves accompany the arteries as they run along both sides of each digit. Proper volar digital nerves serve the sensitive pads of the fingers. Surgical injury is to be avoided; lacerated volar digital nerves should be repaired. A numb fingerpad is an annoying disability. (Dorsal digital nerves from the radial and ulnar nerves are of far less clinical import.) The proper volar digital nerves run in company with the proper digital volar arteries just volar to the dorsal ends of the finger flexion creases.

The most important palmar nerve is the **recurrent branch of the median,** which innervates the three thenar eminence muscles and the first two lumbricals. It branches off the median just after the median exits the carpal tunnel. The recurrent branch arches back over the thenar muscles at around the junction of the proximal and middle thirds of the eminence.

Bones, joints and ligaments of wrist and hand.
There are eight carpal bones in proximal and distal rows in the carpus: From radial to ulnar sides, the proximal row contains scaphoid, lunate, triquetrum and pisiform; distal row contains trapezium, trapezoid, capitate and hamate. The pisiform, hook of the hamate and tubercle of the scaphoid are readily palpated.

Radiocarpal (wrist) joint. This joint is between the radius and scaphoid-lunate. The radius and ulna articulate to allow pronation and supination. The ulna, however, is separated from the ulnar proximal carpals by an articular disc. The joint is held together by strong dorsal and volar radial and ulnar collateral ligaments, extending from radial and ulnar styloids to the scaphoid and triquetrum, and dorsal and volar radiocarpal ligaments. Movements permitted at the wrist joint are flexion, extension, adduction and abduction; circumduction here is also possible.

Intercarpal/midcarpal joint. This permits flexion and extension betwen the two carpal rows. Together with movements of the radiocarpal joint and the supination/ pronation provided by the proximal and distal radioulnar joints, this joint affords a remarkable potential for movement and positioning of the hand.

Carpometacarpal joints. Important here is the joint between the trapezium and metacarpal of the thumb. The saddle-like structure of the joint permits the wide variety of motions of the thumb – flexion, extension, abduction, adduction, rotation and circumduction. The other carpometacarpal joint articulations permit some gliding of the fifth metacarpal, less gliding of the fourth and virtually no motion of the third and second.

Intermetacarpal joints. An important structure is the deep transverse metacarpal ligament, which holds the distal extremities (heads) of the metacarpals together and preserves transverse arching of the hand.

Metacarpophalangeal joints. These joints flex and extend and, with the fingers in extension, abduct, adduct and, thus, circumduct. When in flexion, abduction and adduction are impossible. The collateral ligaments around the joints are lax in extension, but tight in flexion.

Interphalangeal joints. These are strictly hinge joints held together by collateral ligaments. Only flexion and extension are permitted.

Veins of the upper extremity.
In general, the veins accompany the arteries. None of the veins is essential. Superficial and deep venous systems are found in the arm as in the leg. The two areas communicate freely. Any upper extremity vein may be ligated with impunity. Ligation of the axillary vein will produce some distal edema, although collateral channels will readily develop.

An obvious collateral is the **cephalic vein**, a superficial vein originating in the radial side of the hand which, as with the saphenous vein of the lower extremity, remains superficial until its entry into the (deep) axillary vein near the latter's termination. The cephalic vein traverses the interval between deltoid and pectoralis major – the deltopectoral goove. In the forearm and arm, the cephalic vein lies under the skin on the radial side of the extremity.

The **basilic vein** is another superficial vein arising in the hand on the ulnar side. It ascends the ulnar side of the posterior aspect of the forearm. At the elbow, it communicates with the cephalic via the vena mediana cubiti (*cubit* = L. elbow). The vein then continues upwards along the medial border of biceps to the teres major border where it pierces the brachial fascia to join the brachial vein to form the axillary vein. The vena mediana cubiti and the basilic vein run in close proximity to the brachial artery at the elbow. As this is a favorite area for venepuncture, this relationship must be borne in mind.

Brachial veins accompany the brachial artery.

The **axillary vein,** with its companion artery, extends from the lateral border of teres major to the lateral border of the first rib where it becomes the subclavian vein. Tributaries of the axillary vein are the same as those for the accompanying artery except for the cephalic vein.

In contrast to the lower extremity, gravity is not a problem to be overcome to return blood to the heart. Venous circulatory clinical problems are virtually non-existent in the upper extremity. The veins are valved.

Lymphatics of the upper extremity.

Lymph nodes. Lymph node clusters are found in the epitrochlear region just above the medial humeral condyle and, more importantly, in the axilla where lymph from the entire upper extremity must pass. In addition, the axillary nodes filter lymph from the breast, and from the skin and muscles of the anterior and lateral thorax, and back of the neck and shoulder. Axillary nodes number more than 30, with the deeper ones clustered around the axillary vein.

Lymph vessels. Lymphatic trunks ascend, in company with the veins, from the apex of the axilla on both sides to enter the area of the internal jugular-subclavian vein junction.

There may be multiple lymph channels here. On the left side, one or all of the trunks may enter the thoracic duct near its termination.

Innervation of the skin. Be aware of the fact that dermatome areas are highly variable and that nerve areas overlap. Therefore, cutaneous nerve injuries are not significant (except for the digital nerves). However, having a general idea of root dermatomes and cutaneous nerve areas is helpful in the diagnosis of major nerve or root injuries either to support the anatomical diagnosis of the origin of motor defects or to elucidate evidence of nerve injury when motor testing is not possible.

Root levels of the thumb, and middle and little fingers are C6, C7 and C8, respectively.

The medial cord is C8–T1. It supplies the **medial** brachial and antibrachial nerves innervating the **medial** aspects of the arm and forearm. It ends as the ulnar nerve to provide sensation to the ulnar side of the hand.

The lateral cord is C5–7. It produces the **musculocutaneous** nerve and supplies the radial side of the forearm as the **lateral** antibrachial cutaneous nerve.

The posterior cord provides the axillary nerve, which innervates the skin over the deltoid muscle. This proximal muscle and skin area are supplied by C5–6 whereas the remainder of the posterior area of the upper extremity is innervated by posterior branches of the radial nerve: C6 for the **posterolateral** side of the arm, forearm and thumb; C7–8 for the **midposterior** aspect of the upper extremity.

The diagnosis of, suspicion of and results of nerve injury in the upper extremity all depend heavily on an understanding of the anatomy of these nerves and the muscles they serve. A few examples are given here.

The knowledgeable physician realizes that the axillary nerve is at risk in shoulder dislocations, fractures around the humeral head and in prolonged abduction, stretch or pressure (crutches) as it winds about the humeral head. Numbness of the skin over the deltoid area and loss of abduction of the arm will be demonstrable.

The radial nerve is at risk in humeral shaft fractures as it rests in its bony groove in its spiral course around the humerus. Injury results in an inability to extend the wrist – 'wrist drop'.

Fractures around the elbow may injure the ulnar, median or radial nerves (and brachial artery) as the sharp ends of bony fragments may tear through surrounding tissue.

Injury to the deep branch of the radial due to fractures, dislocations or surgery around the radial head will greatly limit thumb function (denervation of extensors and abductor).

Suicide attempts or other laceration of the volar wrist area may injure the ulnar or median nerve. Median loss is manifested by an inability to oppose the thumb and numbness of the radial side of the hand. Ulnar injury here results in hyperextension of the metacarpophalangeal joints (unopposed pull of radial nerve-innervated extensor communis) and hyperflexion of the interphalangeal joints (unopposed median nerve-innervated finger flexors) due to intrinsic muscle paralysis. The ulnar side of the hand will be numb.

The upper extremity is such an important region that mistakes in diagnosis and iatrogenic injury must be avoided.

As skin innervation is so highly variable and overlapping, accurate diagnosis by sensory testing is unreliable. However, in the examination of a badly mangled extremity, sensory testing may be all that is available in physical examination. As a rule, median nerve loss is demonstrable by numbness of the pad of the index finger, ulnar nerve loss by numbness of the pad of the little finger, and radial nerve injury by numbness of the dorsal aspect of the skin web between the thumb and index finger.

Review Exercises

As an anatomical review, sketch in the appropriate contents (muscle groups and compartments, and important vessels and nerves) for each cross-section.

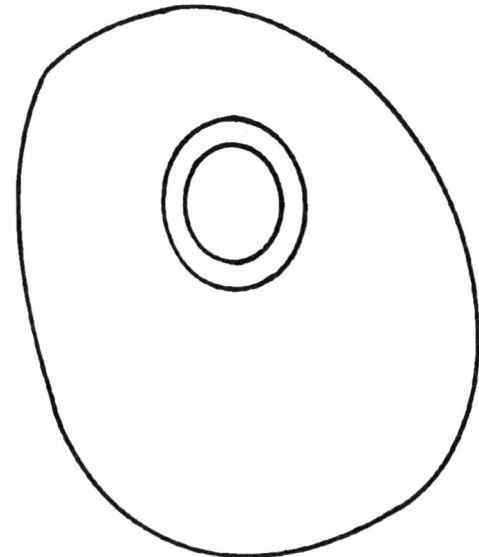

Upper-third of humerus level

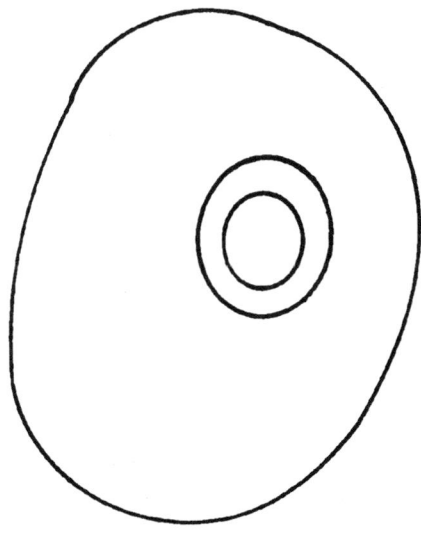

Mid-humerus level

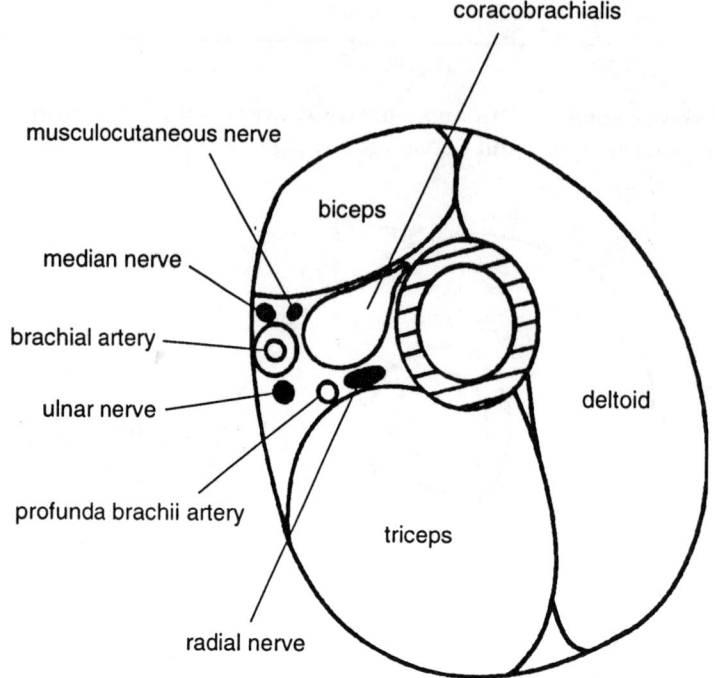

Upper-third of humerus level

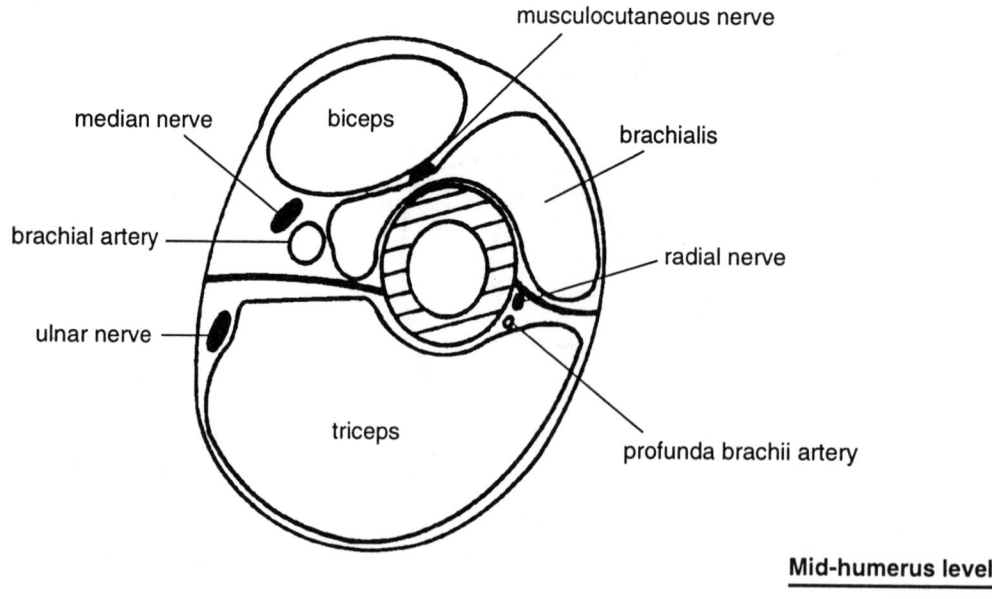

Mid-humerus level

Sketch the contents.

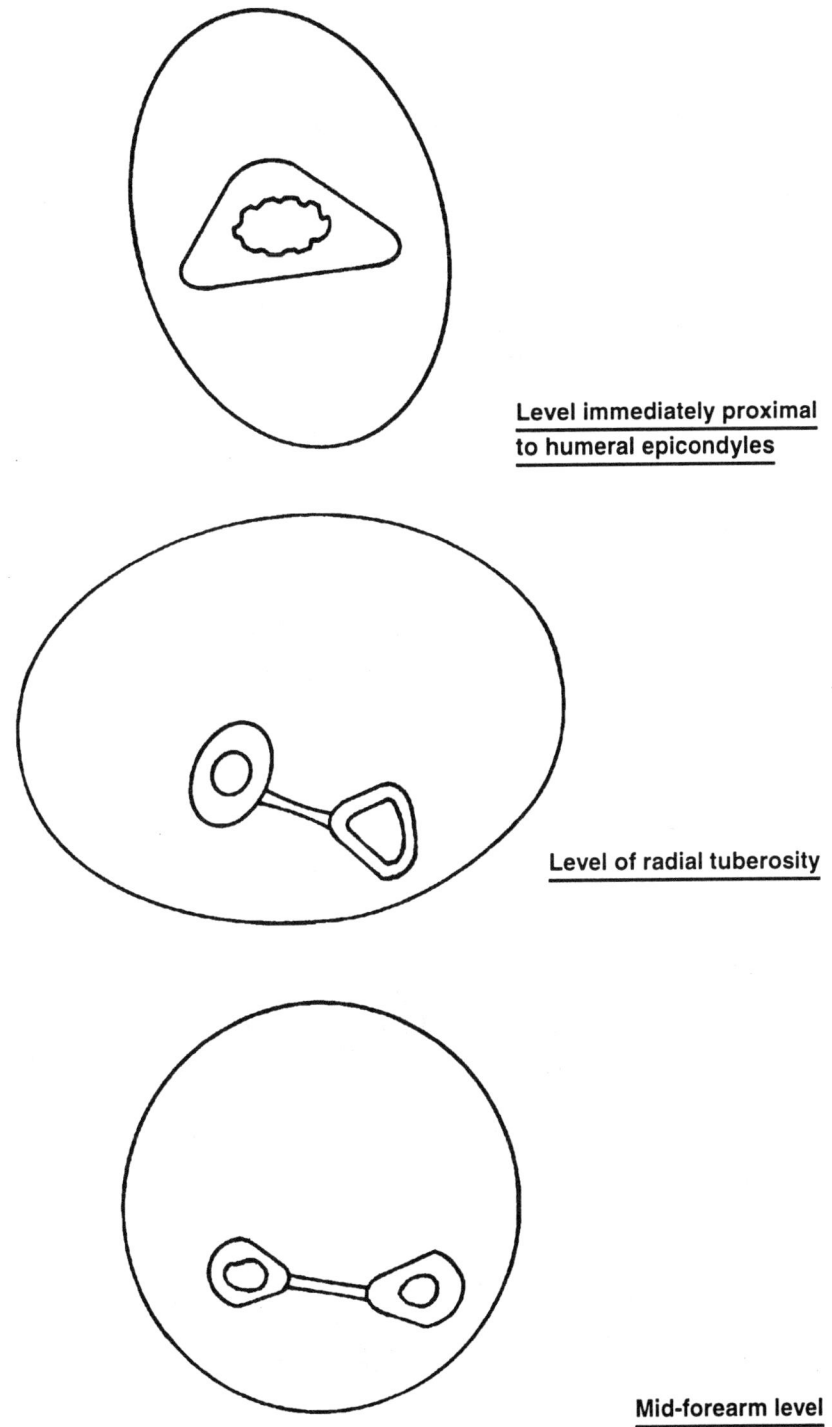

**Level immediately proximal
to humeral epicondyles**

Level of radial tuberosity

Mid-forearm level

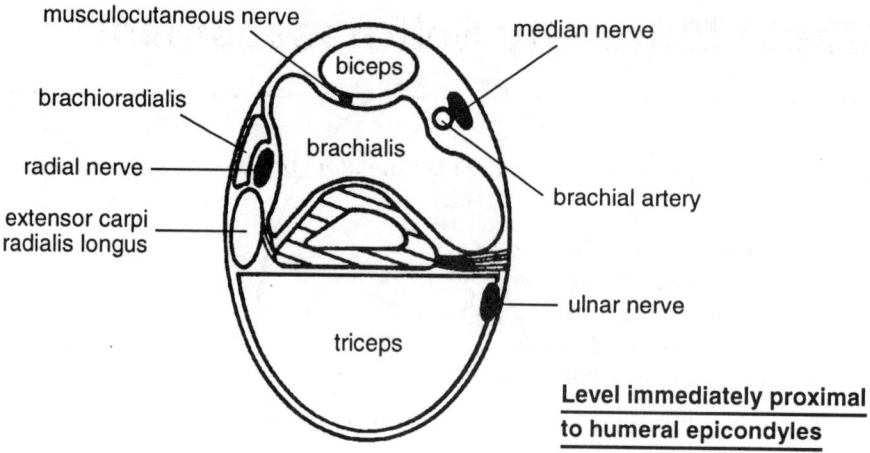

Level immediately proximal
to humeral epicondyles

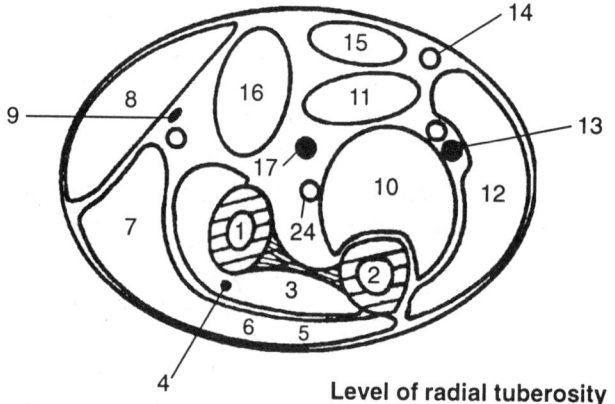

Level of radial tuberosity

1. Radius
2. Ulna
3. Supinator
4. Deep branch of radial nerve
5. Extensor carpi ulnaris
6. Extensor digitorum
7. Extensor carpi radialis longus
   & brevis
8. Brachioradialis
9. Superficial branch of radial
   nerve & radial artery
10. Flexor digitorum profundus
11. Flexor digitorum superficialis
12. Flexor carpi ulnaris
13. Ulnar nerve & artery
14. Palmaris longus
15. Flexor carpi radialis
16. Pronator teres
17. Median nerve
18. Extensor pollicis longus &
    indicis
19. Extensor pollicis brevis
20. Abductor pollicis longus
21. Posterior interosseous nerve
    & artery
22. Anterior interosseous nerve &
    artery
23. Flexor pollicis longus
24. Common interosseous artery

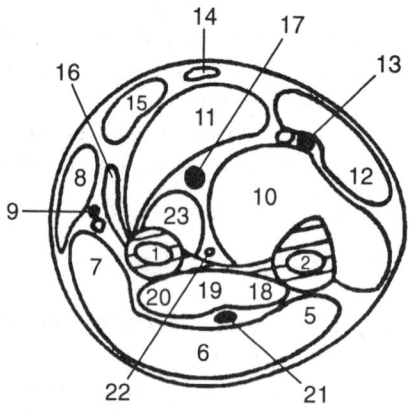

Mid-forearm level

Upper Extremity Self-Assessment

1. The integrity of the acromioclavicular joint depends most importantly on the integrity of the:

    A   Coracoclavicular ligament.
    B   Sternoclavicular ligament.
    C   Acromioclavicular ligament.
    D   Coracoacromial ligament.
    E   Coracohumeral ligament.

2. Regarding the clavicle, choose the one incorrect statement:

    A   Is the most frequently fractured bone of the shoulder girdle.
    B   The sternoclavicular joint is shallow and dislocates frequently with falls on the outstretched upper extremity.
    C   May be considered a strut holding the arm away from the trunk.
    D   Full movement of the upper extremity depends on the free motion permitted by the acromioclavicular and sternoclavicular joints.
    E   In cases of fracture of the clavicle, the subclavius muscle protects the underlying vessels and nerves from injury.

3. Choose the one incorrect ending to the following statement: With injury of the long thoracic nerve, there will be:

    A   Restriction of range of elevation of the scapula.
    B   Restriction of range of upward rotation of the scapula.
    C   'Winging' of the scapula.
    D   Weakness in protraction of the scapula.
    E   Adduction of the arm.

4. Regarding the trapezius muscle, choose the one incorrect statement:

    A   It is innervated by dorsal primary rami of cervical spinal nerves.
    B   It inserts on both the scapular spine and the clavicle.
    C   It is capable of both superior and inferior rotation of the scapula at the acromioclavicular joint.
    D   Embryologically, it is seen as developing as part of the branchial apparatus.
    E   If paralyzed, the ipsilateral shoulder will be lower than the non-paralyzed side.

5. Dislocation of the shoulder:

    A   Is usually accompanied by a tear in the glenohumeral joint capsule in its superior portion.
    B   Is aided by a levering action in abduction, with the humerus as lever and the acromion as fulcrum.
    C   Often injures the musculocutaneous nerve, resulting in weakness in elbow flexion.
    D   Is usually caused by a fall on the outstretched upper extremity.
    E   Results in inflammation of the subacromial bursa.

6. Regarding the so-called 'rotator cuff' of the shoulder, choose the one incorrect statement:

   A   Includes muscles attaching to the greater and lesser tuberosities of the humerus.
   B. Includes four muscles, all innervated by C5, C6, or both.
   C. Is importantly involved in abduction of the arm through its supraspinatus muscle.
   D. Has as its internal rotator the teres minor muscle.
   E. The supraspinatus muscle is protected from injury by the subacromial bursa.

7. Regarding pectoralis major, choose the one incorrect statement:

   A   Is innervated by the underlying long thoracic nerve.
   B   Serves as a medial rotator of the arm along with latissimus dorsi.
   C   If paralyzed, the arm will be in an extended position at rest due to unopposed tonic pull of latissimus dorsi and teres major.
   D   Is of major importance in climbing trees.
   E   Inserts on the crest of the greater tubercle of the humerus.

8. Regarding the axilla, choose the one incorrect statement:

   A   It is a pyramidal space walled in by the pectoral muscles in front, latissimus dorsi, teres major and subscapularis behind, and the chest wall medially. The three meet at an apex at the area of crossing of the clavicle and first rib.
   B   Contains lymphatics running from the upper extremity, chest wall and breast to the internal jugular-subclavian vein junction.
   C   The anterior axillary fold is formed by the pectoralis minor muscle.
   D   Contains the entire blood, lymph and nerve supply for the upper limb.
   E   The brachial plexus and axillary vessels are encased in a fascial sheath derived from the prevertebral cervical fascia.

9. Looking at a cross-section of the midarm:

   A   Three muscles are seen.
   B   The radial and medial nerves cluster around the radial artery.
   C   The arm muscles seen operate the elbow joint only.
   D   All the muscles seen are innervated by posterior cord brachial plexus branches.
   E   The radial artery is seen deep to the humeral shaft.

10. The triceps muscle:

   A   Is homologous to the rectus femoris muscle of the thigh.
   B   Inserts on the radial tubercle.
   C   Originates in part from the coracoid process.
   D   Is innervated by the dorsal primary ramus fibers in the radial nerve.
   E   Due to its insertion and the embryological twisting of the limbs, has a supinator action on the forearm.

11. Supination of the hand:

   A   Depends on the integrity of the deep branch of the radial nerve.
   B   May be accomplished by either the ulnar or radial nerve.
   C   Cannot be accomplished by biceps brachii, which is a member of the flexor/pronator group.
   D   Requires the activity of two joints.
   E   May be accomplished by the extensor carpi ulnaris muscle if supinator muscle function is lost.

12. Regarding the brachial plexus, choose the one incorrect statement:

   A   Is formed from ventral primary rami of C5–8 and T1.
   B   Injury to lower roots of the plexus provokes changes primarily in the distribution of the medial cord.
   C   Injury of the upper roots of the cord (C5–6) produces a limply hanging arm with pronated forearm.
   D   Produces cords which are named for their relationship to the axillary artery.
   E   If destroyed, will totally eliminate shoulder action.

13. With a fracture of the midhumerus which injures a neighboring nerve, the most likely finding will be a loss of:

   A. Flexion at the elbow.
   B. Extension of the elbow.
   C. Flexion at the wrist.
   D. Extension of the wrist.
   E. Supination of the forearm.

14. Select the incorrect statement: Collateral circulation in the upper extremity is afforded by:

   A   Patent connections between vertebral and anterior thoracic arteries.
   B   Connections between circumflex scapular and suprascapular arteries.
   C   Vascular arches connecting radial and ulnar arteries.
   D   Communication between collateral brachial and profunda brachial arteries with the recurrent radial and ulnar arteries.
   E   Communicans in the shoulder between thoracoacromial and circumflex humeral vessels.

15. Select the incorrect statement: Division of the ulnar nerve at the wrist will result in:

   A   Weakness in adduction of the thumb.
   B   Loss of function of flexor carpi ulnaris.
   C   Hyperextension of the metacarpophalangeal joints and hyperflexion of the interphalangeal joints of the ring and little fingers.
   D   Loss of sensation of a variable dimension on the medial side of the hand.
   E   Marked atrophy of the intermetacarpal space between thumb and index finger.

16. Regarding the flexor tendons of the fingers, choose the one incorrect statement:

    A All pass through the carpal tunnel.
    B Superficial flexors insert on the middle phalanges.
    C All digital flexor tendons glide through osseoaponeurotic tunnels lined with a synovium-secreting layer.
    D Lumbrical muscles aid the flexor tendons in flexing the interphalangeal joints.
    E Injury to flexor tendons in the digital tendon sheaths is a difficult clinical problem due to the confined space housing the tendons.

17. Regarding the thumb, choose the one incorrect statement:

    A. Normal function of the thumb depends on the median, ulnar and radial nerves.
    B. Opposition of the thumb depends on activity of the thenar eminence muscles innervated by the recurrent branch of the median nerve.
    C. Skin sensation over the pad of the thumb is furnished largely by digital nerve fibers from C6.
    D. The radial nerve is of least importance in the function of the thumb as it is not involved with the intrinsic muscles of the thumb.
    E. The extensors and long abductor of the thumb are supplied by a nerve which is at risk in operations around the head of the radius.

18. Regarding nerve injury in the upper limb, choose the one correct statement:

    A. Dislocation of the lunate bone may cause compression paralysis of the median nerve.
    B. Injury of the musculocutaneous nerve typically causes difficulty in abducting the arm.
    C. Median nerve injury at the wrist typically causes difficulty in flexing the interphalangeal joints.
    D. Fractures of the humeral shaft often result in injury of the deep branch of the radial nerve.
    E. Dislocation of the shoulder is often accompanied by injury of the long thoracic nerve.

Answers and Explanations

1. **A** The acromioclavicular ligaments around the acromioclavicular joint proper are weak compared with the stout coracoclavicular ligament. The other choices do not support the acromioclavicular joint.

2. **B** Although the sternoclavicular joint is shallow, its binding ligaments are strong and dislocation is rare. The other statements are all importantly correct!

3. **E** The long thoracic nerve innervates serratus anterior, which indeed holds the scapula against the rib cage and helps elevate, protract and upwardly rotate the scapula. Connected as it is, it is difficult to conjure up any adductor effort by serratus muscle. Serratus is a good example of the different directions of pull made possible by use of different portions of a broad muscle.

4. **A** Trapezius is a 'back' muscle, but is not innervated by dorsal primary rami. It is of branchial origin and has its own spinal accessory nerve. The ipsilateral shoulder will droop due to loss of tonic elevator and upward rotational pull of the paralyzed trapezius.

5. **B** Shoulder dislocation usually occurs with forced hyperabduction that levers the head of humerus through the poorly supported inferior part of the joint capsule. At risk is the axillary nerve running immediately under the humeral head. The subacromial bursa is not involved.

6. **D** The internal rotator is subscapularis. All other statements denote important features of the cuff.

7. **A** Pectoralis major is innervated by lateral and medial pectoral nerves. Together with latissimus dorsi and elbow flexors, it pulls the body up from a hanging position, as when climbing trees.

8. **C** Pectoralis major forms the anterior axillary fold. Using the anatomical fact of the axillary sheath, local anesthesia of the upper extremity can be induced by injecting an anesthetic agent into the confines of the sheath. Thus confined, the agent can efficiently bathe all brachial plexus nerves.

9. **A** Triceps, biceps and brachialis comprise the muscles here. By the midarm, the radial nerve is well on its spiral way around the humerus. The radioulnar joint is moved by biceps in its supination action; the other arm muscles have no radioulnar activity. Innervation of biceps and brachialis is by the lateral cord musculocutaneous nerve. The radial artery has not formed at this level.

10. **A** Triceps inserts on the olecranon, arising from the infraglenoid tubercle and the humerus itself. Innervation is by ventral primary rami in the posterior cord radial nerve. It cannot supinate.

11. **D** Proximal and distal radioulnar joints must twirl. Supination by biceps and its musculocutaneous nerve is strong; the supinator muscle and the deep branch of the radial nerve are not required. The ulnar nerve and flexor carpi ulnaris are not involved in supination.

12. **E**   Despite destruction of the brachial plexus, trapezius can still produce shoulder movement. Injury of lower roots of the plexus involves C8 and T1, components of the medial cord. Upper root injury (C5–6) paralyzes rotator cuff muscles, biceps, coracobrachialis, brachialis and deltoid. This produces the arm attitude described. Pronation is due to unopposed pronator teres pull. (Pronator teres has some C7 innervation.)

13. **D**   The radial nerve is primarily at risk. Injured in midarm, the triceps radial branches have already been given off and triceps function is intact. The obvious paralysis will be of the extensor muscles of the wrist – 'wrist drop'.

14. **A**   Vertebral and anterior thoracic connections would be of little help in delivering blood to the upper extremity. All the other connections are useful.

15. **B**   Flexor carpi ulnaris receives its innervation high up in the forearm. All the other statements are typical of distal ulnar nerve injury – clawing of digits 4 and 5, atrophy of the interossei, loss of adduction of the thumb and ulnar side numbness of the hand.

16. **D**   Lumbricals, due to their attachment to the extensor hood and its lateral bands, extend the interphalangeal joints. Injuries to tendons within tight tunnels usually require placement of an intact tendon (graft) through the tunnel after removal of the injured flexor tendons. Good function of an injured sutured tendon within the sheath will not ensue due to formation of adhesions, swelling and scarring, which bind the tendon to the immobile tunnel. The tunnel and its synovial lining are necessary for sliding of the tendon and prevention of 'bowstringing'.

17. **D**   The radial nerve is vitally important for normal thumb performance. Without the long and short extensors, and the long abductor muscles, all innervated by the deep branch of the radial nerve, thumb function is badly crippled. The deep radial branch penetrates the supinator muscle, starting at the level of the head of radius. It must be identified and protected in any operation in the area. All three great nerves listed are necessary for normal function of the thumb (ulnar to the adductor and deep part of flexor pollicis brevis and median to the thenar intrinsics).

18. **A**   The median nerve lies just volar to the lunate. This wedge-shaped carpal bone is squeezed volarwards when dislocated. Median nerve injury at the wrist will not impair function of the already innervated digital flexors. The musculocutaneous nerve innervates muscles flexing the elbow. The radial and axillary nerves are at risk in humeral fractures and shoulder dislocation, respectively.

7 Neck

Bones and joints of the cervical spine. Unique features of the cervical vertebrae are the (usually) bifid ends of their spinous processes (except for C7) and the foramina in their transverse processes which house the vertebral vessels. Otherwise, the cervical vertebrae and their joints demonstrate the usual vertebral features. C1, C2 and C7, however, warrant special attention as they are truly unique.

C7. The large, posteriorly projecting, spinous process that is readily palpable at the posterior base of C7 gives this vertebra its designation of 'vertebra prominens'. This is the only non-bifid cervical spinous process.

The costal element of the C7 transverse process may produce a rudimentary or full-blown cervical rib. Its attachment to the first rib may present a problem for pressure-free passage of the brachial plexus roots and subclavian artery, as these structures must pass through an already (normally) narrow space between the first rib and scalene muscles. Circulatory and/or neuritic symptoms in the upper extremity are not unusual in the presence of cervical ribs.

Finally, the transverse process of C7, although possessing a foramen, usually does not transmit the vertebral artery through this hole. The artery usually passes in front of the transverse process of C7 to enter the transverse foramen of C6.

C1: Atlas. Supporting the head as does Atlas, the Greek Titan who supports the pillars of Heaven on his shoulders, this vertebra has no body, only a ring of bone. During its development, its body fused with that of C2 to form the odontoid process of C2. It has no spinous process either, leaving only a bony ring with large superior articular facets which articulate with the condyles of the occipital bone and circular inferior facets which articulate with the superior facets of the axis. The atlanto-occipital joint permits nodding flexion and extension movements of the head. Circular motion involves the atlanto-axial joint.

The dorsal aspect of the posterior part of the atlantal bony ring contains a groove from the transverse foramen to the margin of its spinal canal. The vertebral artery rests in this support after leaving the transverse foramen on its way to the brain.

C2: Axis. The main feature here is the **odontoid process** or **dens** (L. tooth). Projecting upwards through the ring of the atlas, C2 indeed forms the axis or pivot about which the head and the atlas may rotate.

Atlanto-axial joints. It is important to understand these joints. They are vitally important for keeping the heavy head from sliding about – backwards, forwards, laterally – on top of the spinal column as such movements would endanger the fragile spinal cord. The odonto-axial joint is purely rotatory. The dens is held firmly against the inner curve of the anterior atlas arch by a strong **transverse ligament of the atlas**. Extensions of this ligament upwards to the occipital bone and downwards to the body of the axis give the transverse ligament the shape of a cross; hence the term 'cruciform' is applied. Rotatory movement of the dens against the arch of the atlas is allowed by a synovial joint between the top of the dens and atlas arch anteriorly, and a bursal sac between the dens and cruciform ligament posteriorly. Two **alar ligaments** ('winged') from dens to occipital bone condyles on each side act as check ligaments to control extremes of rotation of the odonto-axial joint. Finally, the name **tectorial membrane** (L. covering) is given to the extension of the posterior longitudinal spinal ligament running from the body of the axis to the occipital bone. (By definition, the posterior longitudinal ligament runs between vertebral bodies and discs. As C1 has no body, there is no disc between C1 and C2; thus, a new name was applied to this 'new' ligament.

The usual strong ligaments bind the lateral articulations of C1 and C2 together. The ligamentum flavum becomes a sturdy **posterior atlanto-occipital membrane (ligament)**. The supraspinous ligament in the neck becomes the thick strong **ligamentum nuchae**, which ends at the external occipital protuberance and helps sustain the weight of the head.

Movers of the head and neck joints.

There are some unique 'extra' muscles moving the head and cervical spine in addition to the 'usual' erector spinae, long flexors (longus colli, longus capitis) and transversospinalis groups.

Suboccipital muscles. These small muscles run from the axis and atlas to the occipital bone, and aid and fine-tune movements of the head.

Splenius cervicis and splenius capitis (Gk. bandage). These flat thin muscles cover and aid the usual extensors and rotators of the cervical spine and head.

Note. Remember that all posterior spinal musculature is innervated by the posterior primary rami of the spinal nerves.

Scalenes (Gk. uneven). These lateral vertebral muscles have unequal sides (as in a scalene triangle). The anterior, middle and posterior scalenes extend from the cervical transverse processes to the first and second ribs. They either lift/fix the ribs, or rotate or laterally flex the cervical spine. Innervation is by branches of the ventral primary rami of the cervical spinal nerves.

Trapezius. Well-known to you as a shoulder shrugger and scapular rotator, due to its attachment to the occiput, ligamentum nuchae and spinous processes, this muscle acting on its own can also laterally flex the neck and simultaneously rotate the head to the opposite side. Acting together, the trapezii extend the neck and head. As an old gill-lifter (branchial arch origin, not somitic), the trapezius nerve supply comes from one of the special branchial nerves, the spinal accessory nerve. This is cranial nerve XI and issues from the jugular foramen in the skull. (This is covered in detail in the section on cranial nerves.) For now, remember that this muscle is not innervated by the usual spinal nerves.

Sternocleidomastoid.   Another old gill-lifter powered by the spinal accessory cranial nerve, its course is apparent from its name. Contraction produces flexion of the neck with rotation of the head to the opposite side. Both muscles contracting together flex the cervical spine. This muscle is an important landmark. It divides the neck into convenient anterior and posterior triangles, and itself covers important neurovascular structures.

Cervical fascia.   There are three layers: investing, prevertebral and visceral.

Investing fascia.   This tough fibrous layer encloses the sternocleidomastoid, trapezius and anterior strap muscles as it surrounds the neck.

Prevertebral fascia.   This deeper fascial sheet binds together all the vertebral muscles as it extends anteriorly over the longus colli and longus capitis muscles in front of the vertebral bodies. Laterally, as we shall see, the ventral rami of the cervical nerves that eventually form the brachial plexus emerge between the anterior and medial scalene muscles, receiving an investment of deep cervical fascia as they penetrate this layer. Similarly, the subclavian artery and vein are invested with tubes of this fascia. The entire nerve and vascular supply to the upper extremity is thus enclosed proximally in a sheet of fascia derived from the prevertebral layer. This was identified as the axillary sheath when we studied the axillary area. (This anatomical feature is useful in local brachial plexus nerve block anesthesia. The sheath confines the anesthetic agent, permitting accurate and extensive anesthesia of the upper extremity with a small volume of anesthetic drug.)

Visceral fascia.   This layer, also termed 'pretracheal fascia', is the fascia of the gut tube. It extends from the base of the skull over the walls of the mouth and pharynx downwards to surround the larynx, thyroid, trachea and esophagus as the latter two disappear into the thorax. It is separate from the other fascial layers, having a space lying posteriorly between this and the prevertebral layer called the retropharyngeal-retroesophageal cleft. Infections secondary to perforations of the posterior pharynx or esophagus are thus afforded an easy route of spread to the mediastinum.

Carotid sheath.   This fascial tube contains the common and internal carotid arteries and internal jugular vessels, and the vagus nerve. It is loosely attached to all three cervical fascial layers and walls off the retroesophageal cleft laterally.

Arteries of the neck.   The head and neck are supplied by two arterial systems: the subclavian and the carotid.

Subclavian artery system.   The subclavian arteries rise into the root of the neck from the brachiocephalic trunk on the right and aortic arch on the left. They arch upward to a level just above the clavicle, then course laterally to pass between anterior and middle scalenes and, finally, over the first ribs to become the axillary arteries. During this course, the subclavians are draped over the cupulae of the pleurae. The roots of the brachial plexus (cervical nerve ventral rami) approach the arteries from above in their passage between the scalene muscles. The subclavian veins pass into the mediastinum by running anterior to the anterior scalene, but more superficially and lower than the accompanying arteries. On the left, the thoracic duct lies medial to the left subclavian artery as it enters the neck. The duct then crosses the vertebral artery and upper border of the subclavian as it turns laterally to proceed more superficially to run behind the carotid sheath, and finally turns medially into the lateral aspect of the subclavian jugular junction.

The anterior scalene conveniently divides the subclavian artery into three parts: the first part extends to the medial border of the scalene; the second is that part lying behind the scalene; and the third part extends from the lateral border of the scalene to the first rib, where the artery becomes the axillary.

Important arteries from the first part. These are the vertebral, thyrocervical trunk and internal thoracic. (The small costocervical trunk is interesting albeit unimportant.) There are classically no other subclavian branches. Frequently, however, the transverse cervical arises from the third part.

Vertebral artery. This first branch of the first part of the subclavian ascends between scalene anterior and longus colli to enter (usually) the transverse foramen of C6 (sometimes C5 or C7). Passing upwards through the transverse foramina, it exits that of the atlas and turns posteromedially to run in the groove in the upper surface of the posterior arch of atlas. At the edge of the spinal canal, it turns upwards to enter the skull through the foramen magnum. Branches of significance in the neck are the radicular branches to cervical nerves, meninges and the cord itself.

Thyrocervical trunk. Typically, three branches arise from this first-part trunk: inferior thyroid, suprascapular, and transverse cervical.

Inferior thyroid artery. This ascends in the deep area of the neck, passing behind the carotid sheath and, on occasions, behind the cervical sympathetic trunk as the latter lies in front of longus colli. The main portion turns medially to supply the thyroid gland. The upward extension of the inferior thyroid is the ascending cervical artery, which makes significant contributions to the blood supply of the spinal cord.

Suprascapular artery. This vessel runs laterally behind the clavicle and sternocleidomastoid at its origin, then passes in front of scalene anterior to reach the superior border of the scapula. At the suprascapular notch, it passes over the little superior transverse scapular ligament (separated from the suprascapular nerve by this ligament) to enter the postscapular region (where you have already seen it). It supplies bone and muscle as it goes.

Transverse cervical artery. This crosses the base of the neck at a level higher than the suprascapular, but at the same depth. At the anterior border of trapezius, it divides into superficial and deep branches (seen previously as they supply trapezius and levator scapular-rhomboid areas, respectively).

Internal thoracic artery. Arising opposite the take-off of the thyrocervical trunk from the inferior aspect of the subclavian, it descends behind the costal cartilages (as you should know).

Costocervical trunk. Arising from the posterior aspect of the subclavian, this supplies the first two intercostal spaces through its highest intercostal branch, and the deep tissues of the neck through its profunda cervicalis branch.

Common and internal carotid arteries. The common carotid artery enters the neck from the brachiocephalic trunk on the right and aortic arch on the left. Enclosed in the carotid sheath of cervical fascia, it ascends to around the level of C4 – upper border of thyroid cartilage where it bifurcates into internal and external carotid arteries. The common and internal carotids are covered in their course by the sternocleidomastoid muscle. Within the carotid sheath, the artery is medial to the internal jugular vein and anterior to the vagus nerve. Medially lie the trachea, esophagus and thyroid; posteriorly the sheath rests on the longus colli muscles and the sympathetic trunk. Neither the common nor the internal carotid have any branches in the neck. The internal carotid continues its cervical course in company with the internal jugular vein, resting on the longus capitis muscle and cervical transverse processes to enter the skull through the carotid canal in the petrous portion of the temporal bone.

At the site of bifurcation of the common carotid, a slight bulge is seen in the terminal common and beginning internal carotid. This is the carotid sinus or bulb. Supplied by a branch of cranial nerve IX (glossopharyngeal), it is a baroreceptor center. In the bifurcation crotch lies a tiny 2–3-mm structure, the carotid body, a chemoreceptor also supplied by cranial nerve IX.

External carotid artery. Its origin as described above, this vessel supplies the structures of the neck from the thyroid level upwards. It lies anteromedial to the common carotid, and escapes the cover of sternocleidomastoid to ascend posterior to the ramus of the mandible. It ends here by dividing into its two terminal branches, the continuing superficial temporal artery and the maxillary artery, which supplies the facial structures. This division takes place at the level of the base of the neck of the mandible.

Branches of the external carotid

Anterior:	Superior thyroid
	Lingual
	Facial
Posterior:	Occipital
	Posterior auricular
	Ascending pharyngeal
Terminal:	Superficial temporal
	Maxillary.

Memorize these branches by remembering the areas and structures of the neck that must be nourished: the three anterior branches supplying viscera, muscles, and skin of anterior neck and face; the posterior branches supplying the skin, muscles and scalp posteriorly; and the two terminal branches supplying the face and anterior scalp. Here, we will only describe the arteries in the neck.

Anterior branches.

Superior thyroid artery. Arising as the first branch of the external carotid, this artery descends to supply the superior pole region of the thyroid gland. It has one major branch, the **superior laryngeal artery**, which enters the larynx through the thyrohyoid membrane (together with the corresponding vein and internal branch of the vagal superior laryngeal nerve).

Lingual artery. Arising opposite the tip of the greater horn of hyoid, this runs forwards above the greater horn of hyoid, is crossed by the hypoglossal nerve and plunges into the musculature of the base of the tongue. Turning upwards and then forwards, it supplies the tongue all the way to its tip.

Facial artery. Arising from a common trunk with the lingual or separately just above it, the artery is initially rather deep, lying behind the stylohyoid and posterior digastric muscles. It continues behind the submandibular gland and emerges into view as it attains the anterosuperior border of the gland. Now lying superficially, it crosses the mandible just anterior to the anterior border of the masseter muscle. Here, its pulse can be readily felt. Together with its accompanying vein (which runs superficial to the submandibular gland), it courses diagonally across the face to the region of the orbit. Although arising in the neck, most of its work involves the facial structures. Its cervical duties include support of the submandibular gland. Small branches ascend into the oropharynx-tonsillar areas and are important sources of post-tonsillectomy bleeding.

Posterior branches.

Occipital artery. This good-sized artery arises opposite the anterior facial branch, and runs upwards behind the posterior digastric to pass between the mastoid process and transverse process of atlas. It then turns backwards to pierce the fascia over the origins of trapezius and sternocleidomastoid and run up the posterior scalp in company with the greater occipital nerve. Its branches supply muscle, scalp, skull bone and meninges during its course.

Posterior auricular artery. A small artery arising high in the neck above stylohyoid and the posterior digastric, it proceeds posteriorly under the parotid and across the styloid process to the area between the cartilaginous ear canal and mastoid process. It has an interesting stylomastoid branch which enters the skull foramen of the same name to help supply structures of the mastoid process and tympanic cavity.

Ascending pharyngeal artery. This small, deeply placed, artery arises near the commencement of the external carotid and ascends alongside the pharynx to the base of the skull. It supplies the pharynx and prevertebral muscles.

Terminal branches.

Superficial temporal artery. The external carotid divides into its two terminal branches deep to the parotid behind the neck of the mandible. The superficial temporal is the ascending terminal branch. It ascends over the posterior part of the zygomatic arch to reach the auricular and temporal regions. As it crosses the arch, its pulse is readily palpable. Its **transverse facial** branch emerges from the parotid to traverse the face just above the parotid duct.

Maxillary artery. The important artery of the deep face runs medially behind the neck of the mandible to enter the infratemporal area (described in the Head section).

Veins of the neck.

Veins of the neck. In general, the cervical veins accompany cervical arteries and venous blood terminates in the brachiocephalic veins. However, there are aspects of this venous anatomy that are unique. Think of cervical veins as two systems, one superficial and the other deep. The superficial system includes the external and anterior jugulars superficial to investing fascia. The deep system includes the subfascial internal jugular (L. neck) and vertebral veins.

Superficial jugular system.

External jugular vein. Readily visible on straining, it originates at the junction of the posterior auricular vein and a branch of the retromandibular or posterior facial ('external carotid') vein and is joined lower down by the occipital vein. It runs superficial to the investing deep fascia from the region of the angle of the jaw and parotid, and across sternocleidomastoid to the area above the midclavicle, where it pierces the investing deep fascia to enter the subclavian vein. (Remember, the subclavian vein extends from the first rib medially in front of scalenus anterior to a junction with the internal jugular in the medial neck root area. It lies posterior to the clavicle and at a lower level than the subclavian artery.) Tributaries to the external jugular correspond to branches of the transverse cervical and suprascapular arteries.

Anterior jugular vein. First identifiable in the region of the hyoid, it runs down the neck superficial to the strap muscles between the anterior midline and medial border of sternocleidomastoid. At the base of the neck, it turns laterally, passing beneath sternocleidomastoid to enter the external jugular at the latter's entrance into the subclavian. The two anterior jugulars are usually connected by a venous 'jugular arch' in front of the trachea in the supramanubrial area. Division of this arch contributes to bleeding during tracheostomy! The superficial system receives blood from superficial areas of the head and neck. It is widely connected to the deep system, especially where the external jugular is formed. Here, the retromandibular vein splits: One part joins with the posterior auricular to form the external jugular; the other joins the facial to form a common facial vein which enters the deeper internal jugular.

Deep jugular system.

Internal jugular vein. This vein appears at the base of the skull as a direct continuation of the sigmoid dural sinus, the largest drainer of brain blood, and leaves the skull via the large jugular foramen between the occipital and temporal bones. The internal jugular descends the neck in the carotid sheath of cervical fascia behind the sternocleidomastoid muscle. Within the sheath, it lies anterolateral to the common and internal carotid arteries. (Both arteries and vein are anterior to the third party in the sheath, the vagus nerve.) At its origin, the jugular is behind the parotid but, as it descends, it is crossed from above downwards by the spinal accessory nerve (cranial nerve IX), posterior belly of digastric, omohyoid and anterior jugular vein, in that order. It terminates behind the gap between the sternal and clavicular heads of sternocleidomastoid by joining the subclavian vein.

On both sides, lymphatic trunks enter the internal jugular at this jugular-subclavian junctional area. On the left, the terminal thoracic duct passes behind the jugular to enter the junction on its lateral aspect. On both sides, lymphatic trunks from the head, neck and upper extremity can also be seen entering the venous system at this point.

The retromandibular vein, in addition to giving a branch to join the postauricular vein to start the external jugular, continues and merges with the facial vein to form a common facial trunk. The latter then splits to communicate widely with both the deep internal and superficial external jugular systems. The bulk of the facial trunk blood enters the deep jugular system.

Vertebral system.

Internal and external vertebral (within and outside of the vertebral canal) venous plexuses extend the entire length of the spine and eventually drain into the vertebral and segmental veins.

It should be noted that there are few, if any, functional valves in the entire cerebral-jugular-internal vertebral venous systems. Valveless communicating veins connect all three through the foramen magnum, jugular foramina, intervertebral foramina, orbits and numerous tiny foramina in the skull. The clinical import of this widespread free communication is apparent when considering hematogenous spread of infection or tumor, internal jugular ligation, or cerebral venous pressure. Venous return from the head and neck depends on gravity and negative intrathoracic pressure. There is no need for a valved muscle-pump system as is so necessary below heart level.

Cervical lymphatics.
All body lymph channels end in the neck, dripping their content into the venous system at the junction of the subclavian and internal jugular veins. Lymph channels are visible bilaterally as they reach this junction. The largest and clinically most important channel is the thoracic duct. Injury of the thoracic duct in the left neck results in a steady outpouring of lymph that requires ligation of the duct proximal (upstream) to site of injury. There is no bleeding back from the venous entry of the duct due to the presence of a lymphatic valve. Draining lymph from the thorax, upper extremity, and head and neck are multiple channels emptying into the junction of the right subclavian and internal jugular. Ligation of the thoracic duct will not produce lymphatic backup as the two (right and left) systems are widely connected.

Thoracic duct. This structure rises into the root of the neck and, leaving the cover of the esophagus, arches upwards and laterally to pass in front of the vertebral artery at its take-off from the subclavian. It then passes behind the common carotid (and carotid sheath), and hooks anteriorly and medially to enter the subclavian-jugular junction on its lateral side.

Cervical lymph nodes. Dozens of lymph nodes stud the cervical lymph channels as they course to their venous destination. Despite the large number of nodes 'everywhere', it is important to have a general idea of their anatomical distribution. An enlarged palpable lymph node or lymph node group is an important regional anatomical signpost that indicates the origin of the process causing the enlargement, for example, infection or tumor. A palpable node in the left neck in the area of the thoracic duct may be the only indication of a silent tumor of the left lung, pancreas, or other organ or tissue in the drainage area of the thoracic duct ('Virchow's node').

Superficial cervical nodes and vessels. These lymphatics accompany the superficial venous system that drains the skin and subcutaneous tissue (anterior and external jugular veins).

Deep cervical nodes and lymphatic trunks. These lymphatics drain the deeper structures of the face, mouth, neck and their viscera. When diseased (large, tender or hard), they may be indicative of a deep-lying tumor or inflammation. The main deep trunks and nodes are clustered around the parent tributaries of the internal jugular (retromandibular and facial veins), large deep tributaries of the internal jugular (with branches of the external maxillary vessels), and tributaries from the neck viscera (salivary glands, pharynx, larynx and thyroid). Clinically important node groups are:

Jugulodigastric 'tonsillar' nodes. These are located around the superior internal jugular at the level of the posterior belly of digastric and drain into:

Deep cervical chain internal jugular nodes, a chain of nodes along the course of the internal jugular vein.

Deep supraclavicular nodes ('scalene nodes'; 'Virchow's node' on the left). These indicate a problem in any area drained by the thoracic duct on the left, and in the right thorax and its content on the right.

Pre- and paralaryngotracheal nodes; parapharyngeal nodes. These deep nodes are often significant features of common laryngeal and thyroid tumors. They drain into the jugular chain.

Lymph channels on the right sometimes join to form a short right lymphatic duct, which then empties into the subclavian–internal jugular junction.

Nerves of the neck.

These are the cervical spinal nerves, cervical plexus, proximal brachial plexus, cervical areas of XI (spinal accessory nerve), X (vagus nerve), XII (hypoglossal), IX (glossopharyngeal), and VII (facial cranial nerve), and the cervical sympathetics.

Cervical spinal nerves. Remember that all spinal nerves demonstrate dorsal and ventral primary rami. The dorsal primary rami are responsible for motor and sensory activities of the back, and the ventral primary rami perform these functions everywhere else. Spinal nerves are made up of dorsal (sensory) and ventral (motor) roots which join to form the mixed spinal nerves (the ventral root also contains sympathetic preganglionics). Nerves exit the spine via the intervertebral foramina. Because the uppermost cervical nerve exits between the occiput and atlas, there are eight cervical nerves, but only seven cervical intervertebral foramina. The dorsal root ganglia containing the cell bodies of sensory fibers of the spinal nerves are situated in the intervertebral foramina just medial to the site of junction of the two roots. Dorsal primary rami are given off promptly and provide sensory innervation to the occiput area and posterior neck, and activate the posterior erector spinae muscles. The largest of these posterior primary rami is the **occipital nerve** from C2.

Cervical plexus. This comprises the four uppermost cervical nerve ventral primary rami after their issue into the neck from the interval between the anterior and middle scalenes. Under cover of sternocleidomastoid, interconnections among the rami form a plexus, whence a sheaf of nerves issues to innervate the skin and musculature of the lateral and anterior neck. The point of emergence of this sheaf is at approximately the midpoint of the posterior border of sternocleidomastoid. A local anesthetic injected along the posterior border of the muscle will effectively anesthetize the anterior neck to permit such procedures as a thyroidectomy.

C3–4 fibers innervate levator scapulae together with the dorsal scapular nerve as described below.

Phrenic nerve. The most important branch of the cervical plexus, derived from C4 with a little help from C3 and C5. The nerve runs behind sternocleidomastoid and crosses the anterior surface of scalenus anterior from lateral to medial before disappearing into the thorax.

Proximal brachial plexus. Ventral primary rami from the lower four cervical nerves (plus T1) make up the roots of the brachial plexus. The roots emerge from the interscalene interval (as did the upper four cervicals), and proceed downwards and laterally over scalenus medius. Just above the third part of the subclavian artery, C5 and C6 join, as do C8 and T1, to form the superior and inferior trunks of the brachial plexus, respectively. C7 alone constitutes the middle trunk of the plexus. All of the trunks then immediately disappear behind the clavicle.

The nerves and roots are bound together by an investing sheath of cervical deep fascia bequeathed to them by the prevertebral layer, as already described.

Three important nerves are given off from this proximal supraclavicular cervical portion of the brachial plexus:

Dorsal scapular nerve. Issuing from the superior trunk and originating from C5, this nerve innervates the rhomboids and contributes to levator scapulae.

Long thoracic nerve.   Arising from C5–7 roots, this nerve passes downwards behind the outgoing brachial plexus and axillary vessels to reach and run down the lateral aspect of the thorax. It lies on and innervates the serratus anterior muscle. As you should know, injury to this nerve results in the 'winged scapula' deformity.

Suprascapular nerve.   Arising from the C5–6 trunk, this nerve courses laterally under the posterior belly of omohyoid and under the anterior border of trapezius. Reaching the dorsal border of the scapula, it seeks out and passes through the suprascapular notch to innervate supra- and infraspinatus muscles.

Cranial nerves in the neck.

Spinal accessory nerve (XI).   This nerve appears in the neck at the jugular foramen, the foramen in the gap between occipital and temporal bones which also transmits the internal jugular vein, and the vagus and glossopharyngeal cranial nerves. The spinal accessory turns posteriorly under the posterior belly of digastric and stylohyoid, passes either in front of or behind the internal jugular, and pierces and innervates the sternocleidomastoid muscle. It emerges from the posterior border of the muscle just above its midposterior point, and continues its oblique downward path under the prevertebral fascia resting on levator scapulae. It enters the lower third of trapezius to innervate that muscle. The spinal accessory may be distinguished from cervical plexus branches issuing from the posterior midpoint of sternocleidomastoid because it pierces, rather than passes under, that muscle and is cranial to other cervical plexus branches. The 'accessory' part of XI relates to the vagus and arises with the vagus in the medulla. The 'spinal' part of XI (to trapezius and sternocleidomastoid) arises in the upper cervical spinal cord, passes up into the skull through the foramen magnum and joins the cranial accessory root. The combined roots then exit the skull through the jugular foramen.

Vagus (X).   The vagus nerve carries fibers essential for the proper sensation and function of the laryngopharynx and larynx (as well as preganglionic parasympathetics to the thoracic and abdominal viscera). The vagus is the posterior element of the three contained in the carotid sheath. It enters the neck through the jugular foramen. As you know, all peripheral nerves carrying sensory fibers contain the cell bodies of these fibers in ganglia outside of the spinal cord. Thus, the vagus, which carries sensory signals from the larynx and pharynx as well as from the lower viscera, likewise has sensory ganglia, the so-called jugular ganglion and nodose ganglion. These rest in the jugular foramen and slightly down into the neck. The nodose ganglion can usually be identified as a fusiform swelling in the extreme proximal part of the cervical vagus.

Together with branches from the glossopharyngeal and spinal accessory nerves, the vagus sends fibers to a **pharyngeal plexus** on the surface of the middle constrictor muscle of the pharynx. From here, motor and sensory fibers from the plexus innervate pharyngeal muscle and mucosa.

Two important branches of the vagus in the neck are the **superior** and **recurrent laryngeal nerves**.

Superior laryngeal nerve.   This branch of the vagus arises high in the neck just below the nodose ganglionic swelling. It runs anteriorly, deep to the carotid on the outer surface of the pharynx, towards the interval between the hyoid and thyroid cartilages. It has two important terminal branches:

Internal branch. The **internal laryngeal nerve** passes through the thyrohyoid membrane with the superior laryngeal artery and vein to report sensation from the laryngopharynx and larynx down to the level of the vocal fold.

External branch. The **external laryngeal nerve** is a tiny nerve which turns downwards from its origin high in the lateral neck and runs on the surface of the inferior constrictor muscle to reach the cricothyroid muscle which it innervates. This is an important muscle as it moves the cricothyroid joint. It tenses or relaxes the vocal cords, which are attached in front to the inside of the arch of the thyroid cartilage and behind to the arytenoid cartilages, which are firmly bound to the laminae of the cricoid. Paralysis of this muscle results in a noticeably weak voice. It is the only laryngeal muscle not activated by the recurrent laryngeal nerve.

Recurrent laryngeal nerve. Recalling the embryology, it will be understood that, although originally the adult recurrent nerve ran directly to the sixth branchial arch tissues, the descent of the heart (and lower aortic arches) from cervical to thoracic level dragged both nerves down with it. The nerves must now arc around the sixth aortic arch to return back up to the sixth branchial arch area; thus, they are recurrent. On the left, the sixth aortic arch becomes the ductus arteriosum and, after birth, the ligamentum arteriosum. The left recurrent nerve arcs around this structure and ascends behind the aortic arch. On the right, the sixth and fifth aortic arches disappear, leaving the right recurrent nerve to arc around the fourth aortic arch, the base of the subclavian artery in adults. Having recurred, both nerves ascend within the tracheoesophageal groove to enter the larynx through the cricothyroid membranes just posterior to the cricothyroid joints. These nerves innervate all of the intrinsic laryngeal muscles except the cricothyroids. Recurrent nerve injury produces vocal cord paralysis.

Cardiac branches may be seen leaving the vagus in the neck bound for the cardiac plexuses. Their origin and course are understandable on recalling the descent of the heart.

Hypoglossal nerve (XII). The nerve of the tongue muscles, it enters the neck through the hypoglossal canal in the occipital bone just medial to the jugular foramen. Passing forwards between the internal jugular and internal carotid, it becomes visible just below the posterior belly of digastric. Continuing forwards, the nerve disappears behind the tendons of the digastric posterior belly and stylohyoid muscles resting on the hyoglossus muscle. Finally, the nerve enters the mouth by running under the free posterior margin of the mylohyoid muscle. In the mouth, it innervates the rest of the tongue muscles.

Facial nerve (XII). The only significant cervical branch of the facial nerve is the **marginal mandibular nerve**, which usually traverses the neck during part of its course and, as its name suggests, runs along the inferior margin of the body of the mandible in its course from the parotid area to the region of the corner of the mouth. It is a slender thread of a nerve, lying superficially just deep to the platysma muscle and crossing superficially to the facial vessels. It dips down into the neck for a variable distance – sometimes 1 cm – at the submandibular level before rising to the corner of the mouth. Damage to the nerve results in paralysis of the muscles of the corner of the mouth and significant oral deformity.

Facial cervical branches innervate the platysma muscle of the neck.

Glossopharyngeal nerve (IX).   Exiting the skull together with cranial nerves X and XI through the jugular foramen, this nerve passes downwards and forwards along the posterior border of the stylopharyngeus muscle (the only muscle it innervates on its own). It crosses stylopharyngeus superficially in the lower portion of that muscle and enters the base of the tongue through the aperture between the superior and middle pharyngeal constrictors. The nerve has important general sensory duties in the oropharynx and over the posterior third of the tongue. In addition, it supplies special sensory innervation to the taste buds over the posterior third of the tongue. Other interesting branches of IX in the neck are sensory branches from the carotid sinus and carotid body, a tympanic branch to the middle ear (which then continues on to the otic ganglion in the head and parotid gland) and contributions to the pharyngeal plexus lying on the surface of the middle constrictor. Sensory ganglia for the nerve (superior and inferior) lie in the jugular foramen. The glossopharyngeal also contributes fibers to the pharyngeal plexus.

Sympathetics in the neck.   Sympathetic contributions to **each** cervical spinal nerve by means of postganglionic gray rami communicantes is the rule in the neck as well as everywhere else in the segmental spinal nerve system. Postganglionic cell bodies are present in the cervical sympathetic trunk. (The sympathetic trunk, as you know, runs from the base of the skull to the coccyx.) In the neck segments of the spinal cord (as well as in the lumbosacral spinal cord segments), there are no preganglionic cell bodies. Preganglionic neurons for the cervical segmental nerves come from T1–3 spinal cord segments. White rami communicantes containing preganglionic fibers supplied by spinal cord segments T1–3 branch off from segmental nerves T1–3 to enter the cervical sympathetic trunk. They ascend within the trunk to appropriate cervical sympathetic trunk ganglia containing postganglionic neurons. Synapse here produces postganglionic gray rami communicantes, which leave the trunk and join the cervical spinal nerves.

Preganglionics to the head must also run up the cervical sympathetic trunk to synapse in the uppermost cervical sympathetic trunk ganglion. Postganglionics to the head must then find their postganglionic way to the glands and smooth muscle of the head by 'hitchhiking' along the appropriate cranial arteries. A 'carotid nerve' from the superior cervical ganglion to the internal carotid artery is regularly seen, carrying postganglionics to the distribution areas of the branches of this vessel, for example, to the structures of the orbit. Thus, injury to the sympathetic trunk at any cervical level produces discernible effects in the head (such as eye and pupil changes, facial sweating or vasomotor changes). (This is covered in detail in the Head section.)

The cervical sympathetic trunk ascends on the surfaces of the longus colli and longus capitis muscles posterior to the fascia of the carotid sheath. Its ganglia are 'condensed' and variable in number. However, there is always a **superior cervical ganglion** at the level of the C1–2 vertebrae. It is approximately 2 cm in length and, from its apex, the carotid nerve can be seen. A variable number of middle cervical ganglia are also seen – perhaps one, two or none. A relatively constant **inferior cervical ganglion** can usually be identified lying at the base of the transverse process of C7. This ganglionic mass is usually connected to the T1 ganglion over the neck of the first rib to form a large '**stellate ganglion**'. The deep-lying stellate is readily seen as its greatest diameter usually exceeds 1 cm. Lying on the neck of the first rib, it is medial to the thyrocervical trunk and posterior to the vertebral artery. Note that postganglionics from the stellate and middle cervical ganglia constitute the sympathetic supply to the entire upper extremity.

Multiple branches from the cervical trunk in addition to the gray rami to segmental cervical nerves run up and down the neck. These include pharyngeal, esophageal, thyroid and cardiac branches as well as connections to the external carotid and vertebral artery 'carriers'.

Cervical viscera and muscles.

Included here is the gut tube in the neck (pharynx and esophagus), the larynx and trachea which budded off the gut tube during embryonic development, the submandibular and parotid glands, and the thyroid and parathyroid glands. Muscles of the tongue are discussed with the head. Similarly, palatal and longitudinal pharyngeal muscles, extending down to the pharynx and involved in the deglutition process, are discussed later.

Cervical pharynx and esophagus.   Only the gut tube (laryngopharynx and cervical esophagus) and airway in the neck are described here. (The oropharynx and nasopharynx are covered in detail in the Head section, where the gut tube is described as a continuum.)

The pharynx forms a muscular funnel into which ingested, chewed and lubricated (saliva) food is swallowed, then passed on into the long, muscular, tubular digestive tract. In the mouth and upper neck, this intake system also serves as the airway. In the laryngopharynx, the airway branches off to pursue its own pulmonary course and function.

The musculature of the pharyngeal funnel consists of three constrictors – superior, middle and inferior. The suprahyoid and infrahyoid muscles aid the constrictors in the propulsion of food, and help protect the airway from ingress of food during the process of swallowing. To understand the anatomy and function of 'visceral' cervical muscle, it is best to start with the hyoid bone and laryngeal cartilages.

Hyoid bone.   This easily identified horseshoe-shaped movable bone just above the thyroid cartilage is suspended from the base of the skull by bilateral stylohyoid ligaments. It offers attachment to muscles involved in swallowing – the suprahyoid and infrahyoid muscles – which are described below.

Laryngeal cartilages.   Only the external aspects of the thyroid and cricoid cartilages are described here. They offer attachment to the pharyngeal constrictors and infrahyoid muscles.

Thyroid cartilage. Displaying the familiar notched 'Adam's apple' in front and extending posterolaterally as two flat laminae or shields, the thyroid cartilage, like the hyoid bone, is horseshoe-shaped in cross-section and open at the back. The posterior borders of the laminae extend upwards and downwards as **superior** and **inferior horns**. These serve as attachment points for the thyrohyoid ligaments above and the cricothyroid joint below. The thyroid is connected to the hyoid by a tough **thyrohyoid membrane.**

Cricoid cartilage. Situated directly below the thyroid cartilage, the interval between the two is covered by a tough **cricothyroid ligament.** The cricoid is signet ring-shaped, with a narrow arch in front connected to a broad single lamina in back. The lamina (the 'signet' part) projects upwards behind the thyroid to partially fill the space between the posterior aspects of the thyroid shield. Inferiorly the cricoid is connected to the uppermost tracheal ring by a ligamentous membrane. Bilateral synovial **cricothyroid joints** between the inferior horns of the thyroid cartilage and lateral aspects of the cricoid between lamina and arch permit a rocking action of the thyroid cartilage on the cricoid, thereby causing a tensing or relaxing of the vocal cords, which run between the two cartilages. Movement of the joints is with the cricothyroid muscles innervated by the external branch of the superior laryngeal nerve.

Suprahyoid muscles. There are four pairs of muscles: the digastric, stylohyoid, geniohyoid and mylohyoid.

Digastric has two bellies, one running from the mastoid process to the hyoid, the other from the hyoid to inside the point of the chin.

Stylohyoid extends from the styloid process to the hyoid along with the posterior belly of digastric.

Geniohyoid lies in the midline as a small ribbon of muscle extending from inside the point of the chin to the hyoid.

Mylohyoid is the muscular floor of the mouth. It arises from the inner surface of the entire extent of the body of the mandible, from the symphysis to the level of the last molar tooth. Fibers pass in a broad sheet on both sides down to the hyoid between the anterior belly of digastric and geniohyoid. The sheets meet in the midline and join in a raphe to form a complete floor for the mouth. The posterior margin of the mylohyoid is free. The submandibular gland duct and the hypoglossal nerve enter the mouth and tongue by passing under this free border.

The innervation of these muscles is confusing (but reflects their embryonic origin). The two posterior muscles (posterior belly of digastric and stylohyoid) are innervated by the nearby facial nerve (nerve of the second pharyngeal arch) which exits the skull between the origins of the two muscles through the aptly named stylomastoid foramen. The anterior belly of digastric and mylohyoid are innervated by a branch of the nerve of the first pharyngeal arch (trigeminal, cranial nerve V), the mylohyoid nerve. The small geniohyoid muscle receives twigs from the C1 spinal nerve hitchhiking on the hypoglossal nerve.

All of these muscles aid in swallowing by elevating the hyoid (as we shall see). With the hyoid fixed by the infrahyoid muscles, the suprahyoids act as jaw openers, not of much importance in humans, but absolutely essential for creatures such as snakes!

Infrahyoid muscles. Again, there are four pairs: sternohyoid, sternothyroid, thyrohyoid and omohyoid. These are the 'strap muscles'.

Sternohyoids are the vertical straps on either side of the midline extending from manubrium to hyoid.

Sternothyroids are the two straps immediately under the sternohyoids and extending from manubrium to thyroid laminae.

Thyrohyoids are continuations of the sternothyroids running from thyroid laminae to hyoid.

Omohyoids (*omos* = Gk. shoulder) have two bellies and extend from the cranial border of the scapula, just medial to the suprascapular notch, to the hyoid. In its midportion, each is held and pulled downwards towards the clavicle and first rib by an extension of the deep cervical fascia so that the upper belly runs up and down adjacent to the sternohyoids and form lateral straps whereas the lower belly runs diagonally across the neck to reach the scapular notch area.

All of these muscles are innervated by branches of the cervical plexus; all depress the thyroid and hyoid cartilages. Although not absolutely essential in humans, these muscles serve as important landmarks for the surgeon. Occasionally, the posterior belly of omohyoid is palpable and mistaken for an enlarged cervical lymph node!

Pharyngeal muscles (*pharynx* = Gk. throat) are found in three sections: the nasopharynx, oropharynx, and laryngopharynx. This description of the pharynx in the neck is confined to the three more-or-less circular constrictors and the three vertical pharyngeal muscles.

Superior constrictor of the pharynx. This originates in front as the posterior continuation of the buccinator muscle of the cheek and hangs from the base of the skull superiorly. A fibrous raphe, the **pterygomandibular raphe**, marks the junction of buccinator and constrictor. The anterior origin of the constrictor therefore extends from the pterygoid hamulus above to the mandible body below at the level of the posterior extremity of mylohyoid. Superiorly the constrictor originates from the medial pterygoid plate above its hamulus and crosses the body of the sphenoid to a bump on the occipital bone just in front of the foramen magnum – the **pharyngeal tubercle**. Muscle fibers curve backwards to meet in a posterior midline raphe. The result is an open-fronted muscular funnel that receives food and air from the mouth, and air from the nose.

Between the skull base (sphenoid body) and top of the muscular sphincter, there is space through which the pharyngotympanic (eustachian) tube and the levator muscle of the palate pass to their respective destinations.

Middle constrictor. This originates from the hyoid bone and stylohyoid ligament. Its upper border overlaps the lower border of the superior constrictor like nested paper cups. Its fibers curve backwards to meet in a continuation of the same posterior midline raphe as used by the superior constrictor.

Through the slit between the superior and middle constrictors pass the glossopharyngeal nerve, stylopharyngeus muscle and stylohyoid ligament.

Inferior constrictor. This arises from the laminae of the thyroid cartilage and the arch bar of the cricoid. Its fibers also curl backwards to meet in a posterior midline raphe. The upper border of the inferior constrictor shingles over the lower border of the middle constrictor.

Through the gap between the middle and lower constrictors run the internal branch of the superior laryngeal nerve and the superior laryngeal artery. Under the lower border of the inferior constrictor run the recurrent laryngeal nerves en route to the laryngeal muscles and mucosa.

Inferiorly the muscle of the inferior constrictor blends with the musculature of the tubular esophagus.

Stylopharyngeus. This vertical, rather than circular, pharyngeal muscle arises from the styloid process and passes down across the superior constrictor. At its termination, it slips between the superior and middle constrictors to insert on the posterior border of the thyroid cartilage. Some of its fibers insert on the pharyngeal wall itself. The glossopharyngeal nerve (cranial nerve IX) accompanies the muscle and innervates it.

Salpingopharyngeus and palatopharyngeus (*salpinx* = Gk. tube) These two small, vertically orientated muscles arise from the orifice of the eustachian tube and the soft palate, respectively. They join below the tonsils with stylopharyngeus, and all proceed downwards to insert on the posterior aspect of the thyroid cartilage and wall of the pharynx.

All of these muscles except stylopharyngeus are innervated by the pharyngeal plexus, which lies on the surface of the middle constrictor and is formed by fibers from cranial nerves IX, X and XI. In general, the circular muscles propel food downwards; the vertical muscles elevate the pharynx (more on deglutition below).

Submandibular gland. One of the three major salivary glands (along with the parotid and the sublingual), it is approximately the size and shape of a walnut. It rests on the hyoid bone, and crotch of the two digastric bellies and the stylohyoid muscle. Superiorly it snuggles up to and slightly under the body of the mandible. On its inner side, it rests on the hyoglossus and mylohyoid muscles. The facial artery grooves its deep surface and the facial vein runs across its superficial surface. (The facial artery and vein meet at the superior margin of the gland and cross over the body of the mandible together to reach the face.) From the deep side of the gland issues a duct which runs forwards under the free border of mylohyoid to place its orifice in the mouth alongside the base of the frenulum of the tongue. Usually, a tongue of glandular tissue encases the proximal part of the duct. Sympathetic postganglionics reach the gland by riding on the facial artery. Parasympathetic preganglionics are delivered by a branch of the facial nerve (chorda tympani) to a **submandibular ganglion** which supplies postganglionics to the gland.

Parotid gland. This, the largest of the salivary glands, occupies the space between the external ear and ramus of mandible. Its duct extends forwards over the upper masseter muscle surrounded in its proximal part by a tongue of parotid glandular tissue. At the anterior border of masseter, it turns medially to pass through buccinator and the mucous membrane of the cheek to end in an oral orifice in the cheek just opposite the upper second molar tooth. The transverse facial artery, a branch of the superficial temporal artery, runs forwards across the cheek just above the duct.

The most important relation of the parotid is the facial nerve (cranial nerve VII) and its branches to the muscles of facial expression. The nerve passes through the parotid before appearing on the face. Cranial nerve VII exits the skull through the stylomastoid foramen directly posterior to the parotid, between the bony mastoid and styloid processes of the skull and just inferior to the bony external auditory canal. The nerve enters the parotid and divides into its branches within the gland. Branches appear at the anterior border of the gland to spread across the face.

Innervation of the parotid is by sympathetic postganglionics from the superior cervical ganglion and by parasympathetics from the glossopharyngeal (cranial nerve IX) and the **otic ganglion** in the infratemporal region.

Thyroid gland. This consists of two lateral cone-shaped lobes connected across the midline by a narrow glandular isthmus, which lies variably across the second, third or fourth tracheal rings. The apices of the lobes overlie the lower aspects of the laminae of the thyroid cartilage. In addition, a normally small conical extension of the thyroid reaching upwards from the isthmus or medial portion of either lobe is often present. This 'pyramidal lobe' represents the lower portion of the path (**thyroglossal duct**) pursued by the thyroid in its migration from its origin at the base of the tongue to its adult cervical position. Other remnants of the thyroglossal duct may be found anywhere along this path in the form of, for example, lingual thyroid tissue or thyroglossal duct cysts. The thyroid is hidden by the sternohyoid and sternothyroid strap muscles, and the anterior border of sternocleidomastoid. It embraces the trachea, lateral cricoid and thyroid cartilages, and esophagus, and is bound to these structures by the pretracheal or visceral layer of cervical fascia. Laterally deep to sternocleidomastoid, the thyroid is in close proximity to the carotid sheath. The blood supply to the thyroid comes from the already described superior thyroid branch of the external carotid and inferior thyroid artery of the thyrocervical trunk. Lymphatic drainage is to the deep jugular chain either directly or via deep paratracheal nodes. Occasionally, the first manifestations of a tumor of the thyroid are enlarged midline pretracheal nodes.

An important relation of the thyroid is the recurrent laryngeal nerve on both sides. The position of these nerves is the same on both sides in the neck region despite their differing original courses. The nerve is found in the tracheoesophageal groove at the neck base, from where it then runs diagonally upwards across the proximal trachea and cricoid, and behind the thyroid lateral lobes, to enter the larynx just posterior to the cricothyroid joint at the inferior horn of the thyroid cartilage. Remember, these nerves innervate all of the intrinsic muscles of the larynx except cricothyroid. No recurrents – no voice!

Parathyroid glands. There are usually two on each side – superior and inferior pairs – originating from the fourth and third pharyngeal pouches, respectively. Each measures around 3–4 mm in diameter and (usually) rests against the posterior surface of the lateral thyroid lobes. They are impossible to identify with certainty in the cadaver and difficult to identify in the living. The adult upper parathyroid pair come from the fourth pouch, the lower pair from the third. The thymus also originates from the third pouch, then descends into the superior mediastinum. The third-pouch parathyroids are dragged down with the thymus to become the lower parathyroids and may be found anywhere along the descending track, even within the substance of the thymus itself.

Larynx. The organ of the voice and the site of separation of the airway from the food passageway, it lies within the region of the laryngopharynx (hypopharynx) which extends from hyoid to trachea. The larynx itself is an anterior offshoot of the laryngopharynx and constitutes its anterior wall. The larynx is surrounded laterally by '**pyriform sinuses**' and posteriorly by the food-passing pharynx. It is equipped with elaborate mechanisms for prevention of the passage of pharyngeal food and fluid contents into the airway it guards (described with deglutition in the Head section). Finally, it contains the mechanisms for the production of sounds of varying pitch and volume.

The larynx consists of jointed cartilages, intrinsic muscles which move these joints, vocal ligaments and folds, and nerves and vessels that supply sensation, power and nourishment.

Laryngeal cartilages.

Thyroid cartilage (already described).

Cricoid cartilage (already described).

Arytenoid cartilages. These small pyramidal cartilages rest, twirl, rock and glide on articular facets on top of the signet portion of the cricoid on either side of the midline. The three-cornered triangular arytenoid base offers one corner as an 'articular process' (facet), and affords the posterior attachment of the vocal cords at a second corner. The cords run from this arytenoidal corner ('vocal process') anteriorly to the inner aspect of the anterior angle of the thyroid cartilage. Small intrinsic laryngeal muscles insert on the third corner ('muscle process') of the triangular arytenoid base and spin the arytenoids in and out, sliding them towards or away from each other, and tipping them forwards or backwards, thus adducting or abducting the cords, fine-tuning their tension and controlling the glottic aperture.

Epiglottis. This is a paddle-shaped cartilage, the 'handle' of which is connected to the inner angle of the thyroid cartilage just below the thyroid notch. The 'paddle' end extends upwards and slightly posteriorly in a diagonal front-to-back plane to lie behind the root of the tongue, and in front of and above the laryngeal airway entrance.

Tiny corniculate and cuneiform cartilages lie in a fold of tissue running from arytenoids to epiglottis and strengthen the aryepiglottic folds (see aryepiglottic fold below). These are only mentioned in passing for the sake of completeness.

Laryngeal membranes and ligaments.   The whole larynx as a unit has to be air- and watertight to function. This is accomplished by:

Cricothyroid membrane ('conus elasticus').   Bridging the gap between cricoid and thyroid, this membrane continues upwards inside the thyroid, then turns medially to form the **vocal ligament** of the vocal cord which extends from the base of the arytenoid to inside the thyroid cartilage (see below for vocal cord structure).

Thyrohyoid membrane.   This old friend stretches from thyroid to hyoid, and is pierced by the superior laryngeal branches of the superior thyroid artery and the internal branch of the superior laryngeal nerve.

Quadrangular membrane.   This membrane stretches from the sides of the epiglottis to the arytenoids, thus closing off the front and sides of the airway tube. The free posterior border of the membrane is called the **aryepiglottic fold**. The lower end of the membrane stretching from the handle end of the epiglottis to the arytenoids on each side is termed the **vestibular fold** or **false cord**. The space between this false cord and the 'true' cord of the conus elasticus is termed the ventricle of the larynx (see below).

Internal structure of the larynx.   The interior of the larynx is divided into vestibule, ventricle and glottis by two folds of mucous membrane covering the true and false cord terminations of the conus elasticus and quadrangular membrane, respectively. The **vestibule** is the proximal part of the larynx to the fold of the false cord. The **ventricle** of the larynx is the space between the two cord folds. The **glottis** is the vocal part of the larynx comprising the true cords and spaces between them, the **rima glottidis** (*rima* = L. cleft).

Ventricular folds (the false cords) are made up of the mucous membrane encasing the thickened lower border of the quadrangular membrane.

Vocal folds (the true cords) are composed of the mucous membrane encasing the thickened upper border of the conus elasticus, the vocal ligament.

The ventricular and vocal folds are a few millimeters apart. The space between them is termed the ventricle of the larynx.

The terminology here is persnickety! The folds are what you see when you inspect the interior of the larynx. The 'working' parts of the cords are the vestibular and vocal ligaments furnished by the two enclosing laryngeal membranes. It is entirely reasonable (and used in common parlance) to refer to the two folds as the false cords and the true cords.

Intrinsic muscles of the larynx. These muscles, by virtue of their attachments to the jointed arytenoid and thyroid cartilages between which the true cords stretch, can exquisitely and finely adjust the tension of the cords as well as the size and the shape of the rima glottidis.

Note that the larynx has two functions: one is to regulate the size of the airway entry and exit to and from the tracheobronchial tree; the other is to produce sound. (Lips, tongue, teeth, etc. make words from out of this sound.)

There are five muscles that need to be well understood. (There are many others, but they are small and can be skipped at this stage of the game.)

Posterior cricoarytenoid. Originating from the posterior surface of the cricoid lamina, this runs upwards and forwards to insert on the muscle process of the arytenoid. Its action is to rotate the arytenoid outwards to widen the glottic slit. It is the only muscle that can do this! Injury of the nerve branch to this muscle produces an abductor paralysis. With bilateral injury, the cords are tightly closed in the midline. Tracheostomy is then an urgent necessity.

Lateral cricoarytenoid. This pair of muscles runs from the anterior arch of the cricoid back to the muscle process of the arytenoid. Its action is thus opposite to that of the posterior cricoarytenoid. This pair closes the rima.

Arytenoideus. Running between the arytenoid cartilages, these contribute to and adjust closure of the rima.

Cricothyroid. Already described as the operator of the cricothyroid joint, this muscle arises from the anterolateral aspect of the cricoid and inserts on the inner upper posterior aspect of the thyroid. Its contraction rocks the thyroid forwards via the cricothyroid joint and thus tightens the vocal cords.

Thyroarytenoid. Arising from the inner surface of the thyroid cartilage back to the muscle process of the arytenoid, its upper part is adherent to the vocal ligament and is called the **vocalis** muscle. It may be considered part of the true cord and fine-tunes it. The remainder of the thyroarytenoid aids in closure of the rima.

Remember that all these muscles are supplied by the vagus nerve and all except the cricothyroids are supplied by the recurrent vagal branches. The cricothyroids are innervated by external branches of the vagal superior laryngeal nerves. The hoarse voice of a paralyzed cord, the weak voice of cricothyroid muscle paralysis and the inability to breathe in or out due to posterior cricoarytenoid paralysis, among others, are familiar problems in clinical medicine which can only be understood by application of anatomical knowledge. Be aware also that an inability to close the glottis eliminates the ability to cough, sneeze, clear the throat or increase intrathoracic pressure to strain. Pulmonary infection is common in this situation – 'aspiration pneumonia'.

Trachea. Surrounded by posteriorly incomplete cartilaginous rings, this extends downwards from the cricoid to disappear into the thorax. Lying immediately in front of the esophagus, it is crossed by the thyroid isthmus and has an intimate relationship with the recurrent laryngeal nerves laterally. Note that its cartilage rings (and the cartilages of the larynx) developed to prevent collapse of the airway in response to negative pressure generated by the process of inspiration.

Esophagus. Food enters the esophagus from hypopharyngeal spaces around the larynx which occupy the anterior part of the larynogopharynx. The spaces lateral to the larynx are the pyriform sinuses and the space posterior to the larynx is bounded by the inferior constrictor. The inferior constrictor extending from the thyroid and cricoid cartilages forms the tubular esophagus once it is free of the branched-off airway. Anteriorly is the trachea, posteriorly the cervical spine and its covering muscles. Between the esophagus with its visceral cervical fascia, and the spine muscles and their prevertebral fascia is the **retroesophageal space**, which extends from the base of the skull to the mediastinum behind the gut tube.

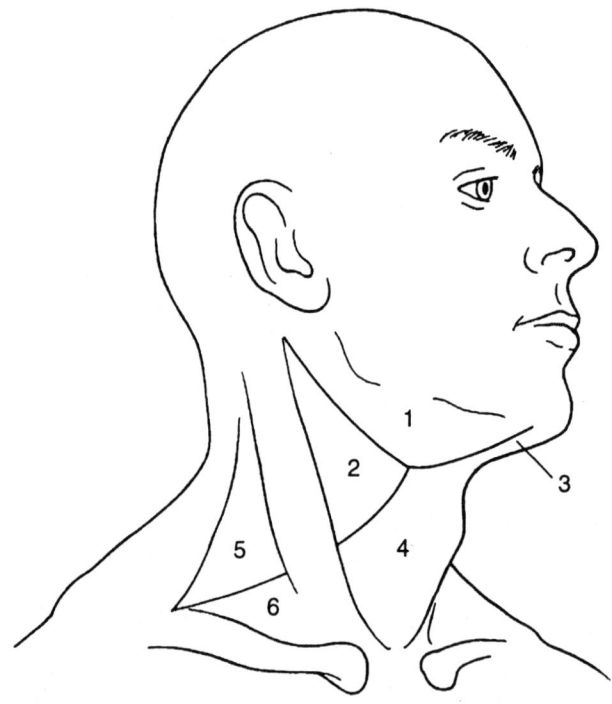

Posterior triangle
5. Occipital triangle
6. Supraclavicular triangle

Anterior triangle
1. Submandibular triangle
2. Carotid triangle
3. Submental triangle
4. Muscular triangle

For organizational purposes, the neck is often divided into triangles as shown above. After studying the anatomy of the neck, you should be able to put any structure in the appropriate triangle.

Neck Self-Assessment

1. On palpating the neck, which of the following is most prominent?

   A Atlas.
   B Cervical rib.
   C Spinous process C7.
   D Odontoid.
   E Spinous process C8.

2. Regarding the atlanto-axial joints, which one of the following is incorrect?

   A The alar ligaments restrict rotatory movements of the head.
   B The nucleus pulposus of the disc between atlas and axis rarely herniates.
   C The odontoid process is jointed with the atlas by means of a synovial joint.
   D The transverse ligament of the axis retains the odontoid process in contact with the anterior arch of the atlas.
   E. The transverse ligament of the axis has vertical extensions to the occipital bone and body of C2, giving the whole a cruciform appearance.

3. Regarding the cervical fascia:

   A The retropharyngeal space is located deep to the prevertebral fascia.
   B Posterior perforation of the esophagus during esophagoscopy results in bacterial seeding of the visceral space enclosed by the pretracheal fascia.
   C The carotid sheath contains the common carotid artery, internal jugular vein, vagus nerve and cervical sympathetic trunk.
   D The axillary fascial sheath is derived from the deep prevertebral fascia.
   E The sternocleidomastoid muscle, being of branchial arch origin, is not included in the investing deep fascia of the neck.

4. In the root of the neck:

   A The subclavian vein passes between anterior and middle scalene muscles.
   B The vertebral artery arises from the most proximal portion of the first part of the axillary artery.
   C The stellate ganglion lies immediately posterior to the neck of the first rib.
   D The thyrocervical trunk of the subclavian artery classically supplies blood to the thyroid, deep tissues of the neck, spinal cord, and superior and posterior aspects of the shoulder.
   E The phrenic nerve passes diagonally across the middle scalene muscle from its lateral to medial border.

5. Regarding the external carotid artery, which one of the following statements is incorrect?

   A Usually has three anterior, three posterior and two terminal branches.
   B Arises at the level of the cricothyroid interval.
   C Its first branch is the superior thyroid artery.
   D Lies deep to the hypoglossal nerve.
   E Originates superior to the inferior thyroid artery.

6. Regarding the lymphatics of the neck, choose the one incorrect statement:

   A The thoracic duct in the neck is in front of the vertebral and subclavian arteries, and behind the common carotid and internal jugular.
   B Enlarged lymph nodes lying on the anterior scalene muscle around the termination of the thoracic duct may indicate disease in the pancreas.
   C The location of enlarged cervical nodes helps locate the site of the disease process causing the enlargement.
   D The thoracic duct may be ligated with impunity.
   E Lymphatic channels other than the thoracic duct are too small to be seen with the naked eye.

7. In performing a radical removal of lymph nodes in the neck, which one of the following would you feel free to remove without fear of severe deficiency?

   A Hypoglossal nerve.
   B Phrenic nerve.
   C Internal jugular vein.
   D Vagus nerve.
   E Recurrent laryngeal nerve.

8. Superficial injection of local anesthesia along the posterior border of the left sternocleidomastoid muscle will produce:

   A Deviation to the left of the protruded tongue.
   B Anesthesia of the entire left anterior neck.
   C Horner's syndrome on the left.
   D Inability to speak normally.
   E Loss of ability to abduct the arm.

9. Regarding the ventral primary rami of cervical nerves C5–8, the following are all correct except:

   A Do not innervate the scalenes as these are supplied by dorsal primary rami.
   B Exit between anterior and middle scalenes.
   C Receive a covering sheath of prevertebral fascia.
   D Before joining to form the brachial plexus, the dorsal scapular and long thoracic nerves are given off.
   E Receive gray rami communicantes from the cervical sympathetic trunk.

Match the numbered item with the most closely related lettered item.

- A Posterior triangle of the neck
- B Marginal mandibular nerve
- C Mylohyoid muscle
- D Posterior cricoarytenoid muscle
- E Cricothyroid muscle
- F Stylopharyngeus muscle
- G Submandibular gland
- H Cleft between middle and inferior constrictors
- I Jugular ganglion

10. Hypoglossal nerve.

11. Vagus trunk.

12. Glossopharyngeal nerve.

13. Spinal accessory nerve.

14. Facial nerve.

15. Recurrent laryngeal nerve.

16. Superior laryngeal nerve.

17. External laryngeal nerve.

18. Lingual nerve.

19. Horner's syndrome (ptosis, miosis, enophthalmus) will result from:

- A Division of the optic nerve.
- B Destruction of the ciliary ganglion.
- C Injury of T1–2 nerve roots.
- D Destruction of the oculomotor nerve.
- E Block of the thoracic sympathetic trunk at T3.

20. Elevation of the larynx is an important part of the deglutition reflex. It is accomplished by several muscles including:

- A Middle constrictor of the pharynx.
- B Palatoglossus muscles.
- C Stylopharyngeus muscle.
- D Levator veli palatini muscle.
- E Styloglossus muscle.

21. Regarding the thyroid, which statement is correct?

    A   The pyramidal lobe is a remnant of the embryonic thyroglossal duct.
    B   The external branch of the superior laryngeal nerve innervates the thyroid and is therefore divided when total thyroidectomy is performed.
    C   If the isthmus of the thyroid lies across the upper four tracheal cartilaginous rings, tracheostomy cannot be done.
    D   The two arteries that supply most of the blood to the thyroid gland are branches of the external carotid artery.
    E   Due to the difference in paths of the two recurrent laryngeal nerves, the left nerve is found lateral to the thyroid and is not as much at risk during thyroid surgery.

22. Regarding the larynx, which statement is incorrect?

    A   The arytenoid cartilages are jointed to the upper margin of the cricoid lamina.
    B   The true vocal cords are attached front and back to elements of the thyroid cartilage.
    C   The posterior cricoarytenoid muscle is the sole opener of the rima glottidis.
    D   The sensitivity of the laryngeal mucosa depends on the integrity of the vagus nerve.
    E   The cricothyroid joint is operated by the cricothyroid muscle, the sole laryngeal muscle not innervated by the recurrent laryngeal nerve.

23. Regarding the salivary glands, which statement is incorrect?

    A   The submandibular gland duct passes under the free border of the mylohyoid muscle.
    B   The facial nerve branches to the muscles of facial expression traverse the substance of the parotid gland.
    C   Innervation of all salivary glands is the responsibility of the chorda tympani.
    D   The parotid duct penetrates the buccinator muscle to reach its oral orifice.
    E   The otic and submandibular ganglia supply postganglionic neurons to the major salivary glands.

Answers and Explanations

1. **C** The 'vertebra prominens'. In fact, T1 may be more prominent, but it is not the first prominence felt going down the neck. (There is no C8!) The other structures are not palpable.

2. **B** There is no disc between atlas and axis. The atlas has no body and thus there can be no disc. The atlas body became the odontoid process of the axis in its embryological history. All other statements are true and important data on the atlanto-axial joint.

3. **D** The retropharyngeal space (continuous with the retroesophageal space) lies between the pretracheal fascia and its visceral content, and the prevertebral fascia. The space extends from the skull base well down into the thorax. The 'favorite' way to infect the space is by perforation of the posterior wall of the esophagus. The carotid sheath does not contain the sympathetic trunk, which lies posterior to it on the prevertebral fascia over longus colli. Sternocleidomastoideus (and trapezius) are enclosed in deep investing cervical fascia.

4. **D** The thyrocervical trunk with its inferior thyroid, ascending cervical, transverse cervical and suprascapular artery branches supplies all these areas. Spinal cord branches pass through the intervertebral foramina as radicular branches nourishing nerve roots and cord. The subclavian vein passes in front of the anterior scalene. The subclavian artery passes between the two scalenes. The stellate ganglion lies immediately anterior to the first rib neck. Vertebral artery is a branch of the subclavian artery. Phrenic nerve passes diagonally across scalenus anterior.

5. **B** Carotid bifurcation is higher, at the level of the upper margin of the thyroid cartilage. All the other statements are correct.

6. **E** Grossly visible lymphatic trunks from head, neck, upper extremities and right thorax are usually present. The abdominal organs drain lymph into the thoracic duct. As enlarged diseased nodes are not palpable until the neck region is reached, palpable nodes here may be the first sign of abdominal disease. Similarly, nodes draining head and neck structures drain specific anatomical areas. For instance, nodes draining a thyroid cancer first drain laterally into deep jugular nodes, not into nodes at the angle of the jaw, etc. Ligation of the thoracic duct is reasonable as collateral channels to the right are present and adequate to handle an added load.

7. **C** Collateral venous channels are adequate to return venous blood from the head and ipsilateral neck without back-up edema. Destruction of any of the other items produces obvious significant disability.

8. **B** Such injection affects the cervical plexus nerves as they escape from under sternocleidomastoid. These nerves supply sensory fibers to the neck structures. The hypoglossal nerve, sympathetic trunk, recurrent laryngeal nerve and C5–6 brachial plexus roots are too deep to be affected. The skin branches will be blocked.

9. **A** The scalenes are not 'back muscles'. They are innervated by branches of ventral

primary rami of cervical nerves. All spinal nerves receive gray rami communicantes from the sympathetic trunk.

10. **C** The nerve passes into the mouth under this muscle.

11. **I**

12. **F** The only muscle innervated by IX (although IX contributes to the pharyngeal plexus).

13. **A**

14. **B**

15. **D**

16. **H** Its internal branch, along with the superior laryngeal vessels.

17. **E** The only laryngeal intrinsic muscle not innervated by the recurrent laryngeal.

18. **G** Carries chorda tympani and its parasympathetic preganglionic fibers destined for the submandibular ganglion, and submandibular and sublingual glands.

19. **C** Horner's syndrome is produced by injury or anesthetic block of the sympathetics to the eye. These originate in T1–2 cord segments. The ciliary ganglion and oculomotor nerve are purveyors of parasympathetics to the eye. The optic nerve contains no sympathetic or parasympathetic fibers. Block of the thoracic sympathetic trunk will not affect higher sympathetics.

20. **C**

21. **A** The thyroglossal duct from the glossal foramen cecum to the adult thyroid may be encased at any point with thyroid tissue it produced. Most frequently, the encasing tissue is seen pointing towards the mouth, attached to the upper border of either the thyroid lobe or the thyroid isthmus, the 'pyramidal lobe'. It is of entirely normal thyroid tissue. The external branch of the superior laryngeal nerve must be carefully avoided during thyroid surgery. It is the only motor nerve supply of the cricothyroid muscle, which is the major tensor of the vocal cords. Significant voice change occurs with its loss. If the isthmus is in the way of tracheostomy, retraction or division is simple and harmless. The inferior thyroid artery is a branch of the subclavian thyrocervical trunk. Despite differing paths at the sites of their recurrence, in the neck the two recurrent nerves are in similar posterior close relationship to the thyroid gland.

22. **B** The anterior extremities of the cords are fixed to the thyroid cartilage, the posterior to the arytenoid cartilages resting on and jointed to the lamina of the cricoid. Sensation in the larynx depends on the vagal superior laryngeal nerve (internal branch) down to the level of the cords and the vagal recurrent laryngeal nerve below the cords.

23. **C** The glossopharyngeal nerve via the otic ganglion innervates the parotid. The chorda tympani of the facial nerve innervates the submandibular and sublingual salivaries via the submandibular ganglion. The facial nerve does indeed traverse the parotid and offers significant difficulty during parotid surgery.

⑧ Head

Scalp. There are five layers:

> **S** kin
> > **C** onnective tissue (fascia)
> > > **A** poneurosis
> > > > **L** oose areolar tissue
> > > > > **P** ericranium.

S–C–A stick together. The aponeurosis is the **galea aponeurotica**, which unites the scalp muscles (frontalis and occipitalis), which move the scalp (wrinkle the forehead, elevate the eyebrows in surprise, horror, delight, etc.). There are vestigial auricular muscles which some people learn to use to wiggle their ears. Loose areolar tissue extends under the entire aponeurosis, enabling it to move over the pericranium, the tightly applied periosteum of the cranial bones.

The blood supply is rich, the vessels carried in the connective tissue layer. Lacerations may bleed heavily, abetted by the inability of the vessels to retract due to fixation in the relatively thick, stiff fascia and subcutaneous tissue. Lacerations through the galea gape due to pull of the frontalis and occipitalis muscles, and thus require suturing. The loose layer is clinically important as infection introduced here may rapidly spread over the entire cranium, infect the cranial bones and, via emissary veins, progress to the meninges and the brain itself. Blood supply is tendered by supraorbital branches of the ophthalmic artery, and the superficial temporal, postauricular and occipital branches of the external carotid.

The nerve supply is from the trigeminal nerve back to the ears (supraorbital V1, zygomaticotemporal V2 and auriculotemporal V3), and from the greater occipital (dorsal primary ramus C2) and lesser occipital (ventral primary ramus C2 from the cervical plexus) behind the ears.

Note: Diagrams depicting the external carotid and maxillary arteries, the trigeminal and facial nerves, and autonomics of the head are found at the end of the Head section.

Skull. The skull is described in the following order:

I.   Cranial vault

II. Base of the skull, internal aspect
   A. Posterior fossa
   B. Middle fossa
   C. Anterior fossa

III. Base of the skull, external aspect

IV. Face
    A. Orbit
    B. Nose
    C. Jaws

This description includes **bony** features, and intimate details involving the ear, nasal cavity, mouth, orbit and face. This section should be read with a skull in hand!

The importance of knowledge of the bones of the skull, and their various foramina, canals and ridges, should be obvious when the necessity of the interpretation of skull films, angiography, computed tomography (CT) and magenetic resonance imaging (MRI); the import of the relation of bones to vital soft tissue structures; the high incidence of symptoms, trauma and disease involving the head; and the need for accurate communication are considered.

Cranial vault. The cranial vault or brain case is made up of occipital, parietals, frontals, squamae of temporal and greater wings of sphenoid. Interesting features of the vault include:

The **lambda**, where the **lambdoid suture** between occipital and parietal bones meets the **sagittal suture** between the two parietals – site of the **posterior fontanelle** in the newborn.

The **bregma**, where the **coronal suture** between the parietal and frontal bones meets the sagittal suture – site of the **anterior fontanelle**.

The **pterion** where parietal, temporal, frontal and sphenoid meet, around 3 cm behind the lateral margin of the orbit. This thin bone overlies the stem and early branches of the middle meningeal artery. Common fractures here may tear the artery, resulting in an extradural hematoma.

Frontal sinuses of varying size are found in the supraorbital frontal area.

Base of the skull, internal aspect. From back to front in the middle, the base is made up of the horizontal portion of the occipital, with its foramen magnum, which meets anteriorly the body of sphenoid, with its sella turcica, which in turn meets the ethmoid, with its cribriform plate and crista galli. Laterally, the skull base from back to front consists of occipital, petrous pyramid of temporal, squama of temporal, greater wing of sphenoid, lesser wing of sphenoid and, finally, the horizontal orbital plate of the frontal. The sharp posterior edge of the lesser wing of sphenoid and sharp anterosuperior ridge of the petrous pyramid demarcate three cranial fossae – anterior, middle and posterior.

Posterior fossa. This fossa houses the cerebellum, brachium pontis and brain stem. The dural **tentorium cerebelli**, attached to the petrous ridge laterally and the groove for the transverse sinuses posteriorly, is its roof. Grooves lodging the **superior sagittal, transverse** and **sigmoid dural sinuses** are well defined. The sigmoid sinuses end in the irregular large **jugular foramina** at the junction of the occipital bone and petrous pyramid of the temporal bone just lateral to the foramen magnum. A smaller groove housing the **superior petrosal dural sinus** is seen just below the sharp superior margin of the petrous pyramid. This groove ends at the junction of the transverse and sigmoid sinus grooves. A small groove for the **inferior petrosal sinus** lies at the anterior junction of the petrous pyramid and middle horizontal part of the occipital, and ends in the jugular foramen. Between the jugular foramina and foramen magnum lie the bilateral bony canals for the transmission of the **hypoglossal nerves**. Immediately inferior to the groove of the superior petrosal sinus is the **inferior auditory meatus** on the posterior face of the petrous pyramid through which pass the **facial** and **statoacoustic nerves**.

Middle cranial fossa. Here rest the temporal lobes of the brain. The bony boundaries are: the midline sphenoid body, and tuberculum and dorsum sellae medially; greater wing of sphenoid and squama of temporal laterally; sharp posterior border of the sphenoid lesser wing anteriorly; and anterior face of the petrous pyramid of the temporal posteriorly. The **sella turcica** sheltering the pituitary gland is in the midline. Small pointed **anterior and posterior clinoid processes** mark the corners of the sella. Just anterior to the sella is a transverse **chiasmatic groove** ending appropriately in the **optic foramina** on each side. The foramina pierce the root of the lesser wing of sphenoid as it arises from the sphenoid body. Just lateral to the optic foramina are the elongated **superior orbital fissures** in the greater sphenoid wing.

The trigeminal nerve is a prominent inhabitant of the middle fossa. A **trigeminal depression (Meckel's cavity)** in the anterior face of the petrous pyramid near its anterior apex houses the large trigeminal sensory ganglion (semilunar and gasserian are other names for this ganglion). Directly in front of the ganglion depression, and separated from it by the foramen lacerum (see below), is an **oval foramen** for passage of the mandibular division of the trigeminal. Approximately 1 cm anterior to the oval foramen is a round foramen (**foramen rotundum**) which transmits the maxillary division of the trigeminal. (The ophthalmic division of cranial nerve V passes out through the superior orbital fissure.) Just posterolateral to the oval foramen is a smaller opening, the **foramen spinosum**, through which the middle meningeal artery passes. (The **sphenoid spine** on the external aspect of the skull base gives a name to the foramen as well as attachment to the **sphenomandibular ligament** and part of the origin of the tensor muscle of the soft palate.) Issuing from the foramen are well-marked readily discernible grooves carrying branches of the **middle meningeal artery** out over the inner surface of the skull vault.

Between the trigeminal ganglion recess and the foramen ovale lies the jagged **foramen lacerum**, cut out from the bone of the apex of the petrous pyramid. This is not a true foramen in that nothing of any size or importance passes through it. In life, the openings are closed by plates of cartilage. The internal carotid artery crosses this foramen as it leaves the internal opening of the true bony carotid canal. Progressing anteriorly, a deep **carotid groove** in the body of the sphenoid on either side of the sellar base is seen. Occupying the groove are the internal carotid artery and cranial nerves III, IV, VI, V1 and V2, all within the confines of the cavernous dural sinus. Following the carotid groove anteriorly, it ascends and forks. The medial fork doubles back on itself under cover of the anterior clinoid, then ends medial to the anterior clinoid. The internal carotid perforates the dura at this point to nourish the brain. The lateral fork proceeds to the superior orbital fissure and optic foramen area for the transmission of orbit-bound nerves through the former, and the ophthalmic artery and optic nerve through the latter. Finally, close inspection of the anterior surface of the petrous pyramid near its apex reveals a thin short groove leading anteromedially from a tiny opening transmitting the greater petrosal nerve, a branch of the facial nerve (**the facial hiatus**). Just inferior to this, a tiny opening transmitting the lesser petrosal nerve (branch of the glossopharyngeal nerve) may be discerned.

Anterior cranial fossa.   Here are found the horizontal orbital plates of the frontal bones (supporting the frontal lobes of the brain), the lesser wing of sphenoid, the body of sphenoid anterior to the chiasmatic groove and the cribriform plate of the ethmoid. The orbital plates form the bony roof of the orbital cavity. Marking the anterior midline of the anterior fossa is the **crista galli** (cock's comb) of the ethmoid to which the dural falx cerebri attaches. There are no significant foramina or other bony features here except the ethmoid **cribriform plate** openings for olfactory nerves, and the small **anterior** and **posterior ethmoid foramina** for transmitting vessels and nerves to the nasal cavity.

Base of the skull, external aspect (mandible excluded).   The posterior half is occipital bone; the anterior half is composed of the hard palate and alveolar processes of the maxillae, horizontal processes of the L-shaped palatine bones, pterygoid processes, undersurface of the greater wings and part of the body of sphenoid, the temporo-mandibular joint fossae, temporal tympanic areas, undersurfaces of the petrous pyramids and the mastoid processes of the temporal bones. The occipital bone is noteworthy for its potent **articular condyles**. Their inner surfaces are roughened for attachment of the alar ligaments. The **foramen magnum** is the obvious central feature. Anterior to the foramen is the **pharyngeal tubercle** from which hangs the superior pharyngeal constrictor. The mastoid, tympanic and petrous undersurfaces of the temporal bone abut the narrow anterior central tongue of the occipital bone.

From the **temporal tympanic area** (surrounding the external auditory meatus) arises the long **styloid process**, which gives attachment to the stylohyoid, styloglossus and stylopharyngeus muscles, and stylohyoid and stylomandibular ligaments. Between the mastoid and styloid processes is the **stylomastoid foramen** for transmitting the facial nerve. The temporal petrous undersurface contains the openings of the **jugular foramen**, **carotid canal** and **foramen lacerum**, from back to front. The **mandibular fossa** of the temporomandibular joint lies lateral to the temporal petrous portion in the posterior beginnings of the temporal zygomatic process. Immediately anterior to the mandibular fossa rises the **articular tubercle** of the temporomandibular joint. Proceeding anteromedially from the mandibular fossa are the **sphenoid spine**, **spinous foramen** and **foramen ovale** (all features of the greater sphenoid wing). The spine gives origin to the **sphenomandibular ligament**; the spinous foramen transmits the middle meningeal artery and the foramen ovale the trigeminal mandibular division. The inferior opening of the **foramen lacerum** is medial to the oval and spinous foramina. In between is a bony **sulcus for the auditory tube** (eustachian tube). In the anterior bony margin of the foramen lacerum is seen the small opening of the **pterygoid canal**, which enables the combined greater and deep petrosal nerves (nerve of the pterygoid canal) to traverse the base of the sphenoid body to reach the pterygopalatine fossa/nasal cavity.

Looking forwards, the **nasal choanae** are seen separated by the thin vomer and flanked by the **pterygoid processes**. Each of the latter consists of medial and lateral **pterygoid plates** of thin bone subtending a **pterygoid fossa**, which gives origin to the medial (internal) pterygoid muscle. From the bottom of the medial plate extends a hook-like process – the **sphenoid hamulus** – around which the tendon of the tensor veli palatini turns to reach the soft palate. The **tensor veli palatini** originates from a shallow longitudinal groove, the **scaphoid fossa**, which extends from the base of the medial pterygoid plate to the sphenoid spine and from the lateral cartilaginous wall of the auditory tube. It is lateral to the **levator veli palatini**, which originates from the medial cartilaginous wall of the auditory tube. The lateral (external) pterygoid muscle arises from the **lateral pterygoid plate**. The plate itself forms part of the medial wall of the infratemporal fossa. The **vomer** extends from the sphenoid body to the midline superior aspect of the entire hard palate. The horizontal plates of the **palatine bones** form about one-fifth of the hard palate, and contain in their lateral extents the **greater palatine foramina** for the descending palatine vessels and nerves. The remainder of the hard palate is formed by the **palatine processes of the maxillae**, with their **alveolar ridges** and sixteen teeth, and the **incisive foramina** behind the incisor teeth which transmit the nasopalatine nerve.

Face. Viewed from the side, the striking feature of the face is the **zygomatic arch** ('cheekbone' of the face) consisting of the zygomatic process of the maxilla and the zygomatic bone. Posteriorly the zygomatic arch is completed by the long slim **zygomatic processes of the temporal and zygomatic bones**. The orbital process of the zygomatic bone curls around the lateral margin of the orbit to meet the bone of the orbital roof, the horizontal orbital plate of the frontal bone. In the lateral orbital wall, the zygoma meets the greater wing of sphenoid. The cheekbone (zygoma and zygomatic process of the maxilla) forms the lower rim of the orbit and is often fractured by direct trauma. Subsequent depression of the inferior supporting rim of the orbit must be corrected or the now unsupported eyeball will also descend, resulting in double vision.

The long lateral arch of the temporal and zygomatic bones covers the junction of the **temporal** and **infratemporal fossae**. The former contains the temporalis muscle running to the mandible whereas the latter has the ptergyoid muscles, maxillary vessel and its branches, and mandibular and maxillary divisions of the trigeminal nerve. The boundaries of the infratemporal fossa are: the maxilla anteriorly; temporomandibular articulation posteriorly; undersurface of the temporal squama and greater wing of sphenoid above; alveolar border of the maxilla below; and lateral pterygoid plate medially. The mandibular nerve and mid-meningeal artery enter and leave through their oval and spinous foramina, respectively, in the sphenoid roof of the fossa. The inferior orbital fissure is open to the fossa. This fossa is a busy and important place!

Following the maxilla posteriorly, it meets the lateral plate of the pterygoid process in an obvious deep crack – the **pterygomaxillary fissure**. Following this fissure upwards, it expands into a smallish space – the **pterygopalatine fossa**. At the superior medial extremity of the fossa is an opening, the **sphenopalatine foramen**, which allows communication between the 'outside' pterygomaxillary fissure and infratemporal fossa and the 'inside' nasal cavity. The foramen is principally a notch in the upper end of the vertical portion of the palatine bone. The maxillary artery and vein pass up the fissure into the fossa, then through the foramen into the nasal cavity as the **sphenopalatine artery and vein**. Housed in the pterygopalatine fossa is the **pterygopalatine parasympathetic ganglion**, fed by the **nerve of the pterygoid canal**. The trigeminal maxillary division – V2 – enters the top of the fossa via its foramen rotundum. (The pterygopalatine fossa and its content are described in detail in the description of the fossa below.) The term 'palatine' is used instead of 'maxillary' as the fossa and foramen are bounded medially by the vertical process of the palatine bone, which is insinuated there between the maxilla and sphenoid. The sphenopalatine foramen itself is a notch in the palatine bone. Remember that the palatine is L-shaped. The horizontal part forms the posterior portion (one-fifth) of the hard palate; the vertical part extends upwards. The fossa is formed by an anterior 'recession' of the maxilla; the palatine then becomes prominent as the anteromedial boundary.

Viewed from the front, the facial orbits, nasal opening, and upper and lower jaws are the prominent features.

Orbit. Sturdy orbital rims are surmounted by thick frontal bone, the **superciliary arches**, joined in the midline by a smooth elevated **glabella** (L. smooth). The **superior orbital rim** is frontal bone, pierced or cleft in its midportion by a **supraorbital foramen** (or open cleft) for the passage of supraorbital sensory nerves (trigeminal) of the forehead. The **lateral orbital rim** is the orbital process of the zygoma with its small **zygomaticofacial foramen** for passage of the nerve of the same name to the skin of the cheek. The medial rim is the frontal process of the maxilla. The inferior rim is the orbital process of the maxilla, which contains the **infraorbital foramen** through which the infraorbital nerve passes to reach the skin of the front of the face and the maxillary process of the zygomatic bone. Blood vessels accompany the nerves through all of these foramina. Looking into the orbit, four walls are seen tapering up and inward to an apex.

The **roof** of the orbit is the orbital horizontal plate of the frontal bone. In the lateral extremity of the roof is a bony depression which houses the lacrimal gland, the **lacrimal fossa**.

The orbital **floor** is formed by the orbital processes of the maxilla and zygoma. It is rather thin and also constitutes the roof of the maxillary sinus. At its medial extremity, the maxillary floor meets the tiny lacrimal bone of the medial orbital wall. Running forwards near the middle of the floor is the **infraorbital groove**, connected to the infraorbital foramen by an infraorbital tunnel containing the infraorbital nerve and vessels. At the posterior extremity of the groove and marking the junction between floor and medial wall lies the **inferior orbital fissure.**

The orbital **medial wall** consists of, from front to back, the frontal process of the maxilla, lacrimal bone and the paper-thin lateral lamina of the ethmoid. The lacrimal bone contains the **fossa for the lacrimal sac.** From the sac, the **nasolacrimal duct** descends to the inferior meatus of the nasal cavity. Two small foramina in the lateral aspect of the ethmoid afford passage of the anterior ethmoid vessels and nasociliary nerve, and posterior ethmoid nerve and vessels, respectively.

The **lateral wall** is formed anteriorly by the orbital process of the zygomatic bone and posteriorly by the undersurface of the greater wing of sphenoid. Between the orbital lateral wall and the floor lies the **inferior orbital fissure**, housing the trigeminal maxillary nerve and its branches, and the infraorbital vessels. (The orbital groove extending anteromedially from the fissure across the posterior orbital floor carries the infraorbital nerve and vessels from the fissure to the foramen.) The lateral wall lies between the orbit and temporal fossa.

The **orbital apex** is the lesser wing of sphenoid perforated by the **optic foramen** that carries the optic nerve and ophthalmic artery. Just laterally between the two sphenoid wings lies the **superior orbital fissure**. This opening communicates at the floor with the inferior orbital fissure. The superior fissure transmits cranial nerves III, IV, VI and V1.

Nasal opening. **Nasal bones** wedged up between the frontal processes of the maxillae support the nasal bridge. (Posteriorly, these maxillary processes form the medial orbital rim.) The remaining nasal structures are internal and are described with the nasal cavity.

Jaws. The upper jaw is the **maxilla**, with its inferior margin bearing the sixteen upper teeth. The lower jaw, the **mandible**, consists of a horizontal **body** and a nearly perpendicular **ramus**. The angle where the two meet is the **angle** of the jaw which is roughened for attachment of the masseter and medial pterygoid muscles. The ramus ends superiorly in two processes, the **condyle** and **coronoid** process, separated by a semicircular notch, the **mandibular notch**. The coronoid process serves as the point of attachment of the temporalis muscle. The condyle articulates with the temporal mandibular fossa. The medial surface of the ramus is marked in its midportion by a sharp spine, the **mandibular lingula**, for attachment of the sphenomandibular ligament. The spine guards a **mandibular foramen** for passage of the inferior alveolar nerve and vessels. The body demonstrates a ridge on its inner surface for the origin of the mylohyoid muscle, and a foramen on the outer surface near the point of the jaw (**mental foramen** and **mental protuberance**) for the exit of mental and labial branches of the inferior alveolar nerve and vessels.

Interior of the cranium.

Meninges: Dura mater. This is the tough cover of the central nervous system. It is tightly applied to the inner aspect of the bony cranium (in contrast to the loose attachment of the dura of the spinal cord to the bones of the vertebral canal). There is therefore a **potential** epidural space within the cranium. This may become an actual space with tears of the meningeal arteries residing in the grooves of the calvaria (epidural hematoma).

Two significant processes of the dura are the **falx cerebri** and **tentorium cerebelli**. The falx extends downwards between the cerebral hemispheres from the crista galli of the ethmoid to the internal occipital protuberance. The tentorium roofs over the cerebellum in the posterior cranial fossa and supports the occipital lobes of the cerebrum. Attachment of the tentorium is to the posterior clinoid processes anteriorly, the groove in the sharp superior margin of the petrous portion of the temporal bone (housing the superior petrosal dural sinus) laterally and the bony groove for the transverse dural sinus posteriorly. Anteriorly between the clinoids is an oval 'defect' in the tentorium – the **tentorial notch** or incisura – for passage of the brain stem, and communicating nerve bundles to and from the middle and anterior cranial fossae areas.

The dura contains endothelial-lined valveless venous sinuses which return venous blood from the brain to the internal jugular vein via the jugular foramen of the occipital bone. The sinuses are located principally at the sites of attachment of falx and tentorium, and at the free margin of the former. The **superior sagittal sinus** runs along the area of attachment of the falx to the skull. The **inferior sagittal sinus** courses along the free border of the falx to its junction with the tentorium, then posteriorly as the **straight sinus** along this junction to meet and join the termination of the superior sagittal sinus. From this sinus junction ('confluence of sinuses'), **transverse sinuses** run in grooves in the occipital bone to the posterior extremities of the petrous portion of the temporal bones. Here they turn downwards as the S-shaped **sigmoid sinuses** to pass through the jugular foramina.

Other important sinuses are the superior and inferior petrosal sinuses, and the cavernous sinus. The **superior petrosal** is located in a groove along the superior ridge of the petrous pyramid (site of attachment of the tentorium), and meets and enters the termination of the transverse sinus. The **inferior petrosal** runs in the junctional area of the petrous and occipital bones, and passes directly down to the jugular foramen. The **cavernous sinus** abuts the lateral aspect of the sphenoid body. It has already been described as containing within its 'caverns' the internal carotid artery and cranial nerves III, IV, VI, V1 and V2. It reaches the jugular via the petrosal sinuses.

The dural sinuses communicate with the face via the ophthalmic vein terminal tributaries (inferior and superior orbital veins), and branches of the facial, pterygoid and pharyngeal venous plexuses; with the scalp via the emissary veins; and with the vertebral venous plexuses via a basilar venous plexus over the anterior extremity of the occipital bone near the foramen magnum. Blood reaches the dural sinuses from multiple small cerebral veins and from the great cerebral vein of Galen, draining internal brain areas and ending in the straight sinus. There are no valves in these veins.

The nerve supply of the dura is furnished by branches of the trigeminal nerve. The dura is the only pain-sensitive meninx. Dural blood is furnished by meningeal arteries, principally the middle meningeal branch of the maxillary artery.

Arachnoid mater. This is the thin watertight membrane closely applied to the dura. Here again, there is only a potential space between the two, the **subdural space.** Cerebral veins must traverse this potential space to gain access to the dural sinuses. Traumatic shaking of the brain may rupture these dural tributaries. Subsequent venous bleeding may slowly open up the potential space, forming a subdural hematoma.

Deep to the arachnoid is the pia-covered brain. Between arachnoid and pia is an actual space – the **subarachnoid space** – containing the **cerebrospinal fluid** (CSF), formed within the brain ventricular system through capillary-rich choroid plexuses and from pia-covered capillaries within the brain itself. The CSF escapes to the subarachnoid space around the brain through openings in the medulla of the brain stem. The arachnoid does not dip into the sulci and fissures, and irregularities on the brain surface, but bridges over them, forming CSF-filled cisterns of varying size. The larger are named according to their positions, for example, cerebellomedullary or interpeduncular.

Along the course of the superior sagittal sinus are the **arachnoidal villi** (arachnoid granulations), tuft-like projections of arachnoid through the dura into the superior sagittal sinus. In this position, the tufts or villi are covered only by arachnoid and sinus endothelium. It is in these arachnoid villi that CSF passes back into the bloodstream. The arachnoid villi are grossly visible in venous lacunae lying alongside and connected to the superior sagittal sinus. They rest in shallow depressions in the calvaria (foveolae granulares).

Pia mater. This delicate membrane covers the entire CNS. In the brain, it dips into its sulci and fissures, carrying blood vessels into the brain substance. Having passed over the cerebral hemispheres, the pia mater covers the medulla and forebrain, forming vascular roofs rich with blood vessels – the choroid plexuses – where CSF is formed.

Brain. The vast bulk of the brain consists of two cerebral hemispheres, separated in the midsagittal plane by the dural falx lying in a **longitudinal cerebral fissure**. Two ear-like projections on either side represent the poles of the **temporal lobes**, resting in the middle cranial fossae. The temporal lobes are separated from the rest of the hemispheres by a deep **lateral fissure**. The poles of the **frontal lobes** rest in the anterior cranial fossae, the poles of the **occipital** on the tentorium cerebelli. The frontal and parietal cerebral lobes are demarcated by a **central fissure** running transversely across the domes of each hemisphere.

The spinal cord enters the head through the foramen magnum to become the proximal part of the brain stem – the medulla oblongata. The brain stem proceeds upwards through the tentorial notch to bury itself in the base of the hemispheres. Perched atop the lower stem is the tennis ball-sized cerebellum; its lobes are connected ventrally from side to side by a broad bridge, the cerebellar brachium **pontis**. Openings connecting the CSF-filled subarachnoid space with the ventricle of the medulla include: one in the midline under the posterior aspect of the cerebellum – the **foramen of Magendie**; and two laterally, one on each side where the brain stem is at its maximum width – the **foramina of Luschka**. Further forward, the **pituitary** is ensconced in its sella turcica, covered with a dural 'drumhead' (diaphragma sellae) and connected to the brain by a stalk – the **infundibulum**.

Brain blood supply. Entering through the foramen magnum are the vertebral arteries. Farther forward, the internal carotid arteries, having entered and traversed the carotid canal in the petrous temporal bone, foramen lacerum and cavernous sinus and having given off the ophthalmic artery at the optic foramen, double back under the anterior clinoids to pierce finally the dura and enter the cranial cavity just medial to the anterior clinoids. The carotids end in large **anterior cerebral** branches between the two hemispheres and in **middle cerebral arteries** running laterally in the lateral cerebral fissure. The vertebrals show paired **anterior and posterior spinal arteries** soon after entering the cranial cavity. The posterior spinal arteries remain distinct throughout the spinal cord whereas the anterior spinals join at the foramen magnum level to form a single anterior spinal artery. The larger **posterior inferior cerebellar arteries** (PICA) are given off just after the spinals. The two vertebrals join at the lower border of the brachium pontis to form a single **basilar artery**. This vessel gives off large **anterior inferior cerebellar arteries** (AICA) at the caudal border of the brachium pontis before ending in larger **posterior cerebral arteries** at the cranial border of the brachium pontis. These pass laterally over the cerebellum to the posterior aspects of the cerebral hemispheres. Just before terminating in the posterior cerebrals, the basilar emits two large **superior cerebellar arteries**. The posterior cerebral and the superior cerebellar arteries embrace the exiting oculomotor nerves.

The internal carotid and vertebral systems are joined by **posterior communicating artery** branches of the intracranial internal carotids which meet similar branches from the posterior cerebrals. Anteriorly there is an **anterior communicating artery** between the two anterior cerebral arteries. Thus, an arterial anastomotic circle is formed about the ventral surface of the most cranial part of the brain stem, offering collateral help if input from either system is curtailed.

Cranial nerves.   There are 12 cranial nerves:

I	Olfactory nerve
II	Optic nerve
III	Oculomotor nerve
IV	Trochlear nerve
V	Trigeminal nerve
VI	Abducens nerve
VII	Facial nerve
VIII	Vestibulocochlear (statoacoustic nerve)
IX	Glossopharyngeal nerve
X	Vagus nerve
XI	Spinal accessory nerve
XII	Hypoglossal nerve.

Olfactory bulbs and tracts lie along the undersurface of the frontal lobes and over the cribriform plate of the ethmoid.

Optic nerves issue from the optic foramina and (partially) cross in the chiasmatic groove (optic chiasm) just in front of the pituitary infundibulum.

The oculomotor nerve exits the ventral side of the brain stem between the superior cerebellar and posterior cerebral arteries. It promptly enters the cavernous sinus and exits the skull via the superior orbital fissure.

The trochlear nerve is the only cranial nerve to issue from the dorsal aspect of the brain stem. It is tiny; it curls around the brain stem to reach the ventral aspect, then angles forwards to enter the cavernous sinus alongside the oculomotor nerve. It, too, exits the skull via the superior orbital fissure.

Trigeminal is the largest cranial nerve (the large olfactory and optic 'nerves' are in fact brain tracts). It leaves the brain stem in the lateral aspect of the brachium pontis. Its sensory ganglion lies in Meckel's cavity; its V3 branch leaves the skull through the foramen ovale, and the other two – V2 and V1 – pass through the cavernous sinus and exit via the foramen rotundum and the superior orbital fissure, respectively.

Abducens appears near the ventral midline at the junction of the brachium pontis and the brain stem. It passes upwards through the tentorial notch to the cavernous sinus and from there to the superior oribital fissure.

The facial and vestibulocochlear nerves exit more laterally from the same brachium–brain stem junction and run laterally to enter the internal acoustic meatus in the medial face of the temporal petrous pyramid.

The glossopharyngeal and vagus nerves emerge from the lateral medulla in a series of rootlets and leave the skull through the jugular foramen.

The spinal portion of the spinal accessory nerve arises from the first five segments of the cervical cord and ascends along the lateral aspect of the cord, passes through the foramen magnum, and joins the vagus and glossopharyngeal to exit through the jugular foramen. The 'accessory' part of the spinal accessory nerve may be considered part of – an 'accessory to' – the vagus. It forms the lower rootlets of the vagus origin.

The hypoglossal nerve leaves the lower medulla laterally in a series of rootlets and enters the hypoglossal canal in the occipital bone.

In the anterior face of the petrous pyramid, the **greater petrosal nerve**, a preganglionic parasympathetic branch of the facial nerve, exits the facial hiatus. It proceeds anteriorly under the trigeminal ganglion and across the foramen lacerum to enter the orifice of the **pterygoid canal** at the base of the sphenoid body en route to the pterygopalatine fossa of the maxilla. Postganglionic fibers from the superior cervical ganglion riding on the internal carotid artery drop off the carotid to join the greater petrosal nerve as the carotid passes medial to the facial hiatus. The sympathetic fibers are termed the **deep petrosal nerve**. These two petrosals make up the **nerve of the pterygoid canal**, which furnishes autonomic supply to the nasal mucosa and lacrimal gland (after the greater petrosal fibers synapse in the sphenopalatine ganglion housed in the pterygopalatine fossa).

Just ventral to the petrous facial hiatus is a tiny opening transmitting the lesser petrosal branch of the glossopharyngeal which carries preganglionic parasympathetic fibers for the parotid gland. The tiny nerve passes through the foramen ovale to synapse in the otic ganglion in the infratemporal fossa, thence to the parotid gland.

Soft tissues of the face.
The superficial structures of the face, specifically, the nose, eyes, lips, skin nerves and muscles of facial expression, and their nerve and blood supplies, are described first, and followed by the deeper areas of the face.

Lips. These have skin on the outside and a mucus-producing lining on the inside, with the two separated by fibers of the orbicularis oris muscle (see below). The marginal lip skin is unique: it is thin, devoid of keratin, glands and hair follicles, and contains abundant capillaries – hence the smooth red surface and 'vermilion border'. Mucoserous **labial glands** are palpable on the internal surfaces of the lips. Superior and inferior labial branches of the facial artery running at the level of the 'red line' supply blood to the area.

Nose. Nasal cartilages extend down from the nasal bones laterally and, together with a central septal cartilage, maintain the flexibility and shape of the external nose.

Eyes. The lids (**palpebrae**) meet at medial and lateral angles (**canthi**). **Tarsal plates** (*tarsus* = L. broad flat surface) of dense fibrous connective tissue stiffen both lids. Eyelashes are **cilia**. **Tarsal glands** (meibomian glands, named after Meibom, a German anatomist, 1638–1700) are visible as strings of tiny swellings on the inner surface of each lid. Their tiny openings can be seen along the lid margins. They secrete sebaceous material which serves to contain (prevent the overflow) of tears. Small sebaceous glands are associated with the cilia as with hair elsewhere. Obstruction of a tarsal gland results in a cystic **chalazion** under the conjunctiva away from the lid margin which may become infected. Infection of the ciliary sebaceous glands produces a stye or **hordeolum** at the lid margin. This is not the same as a chalazion.

The lining of the lids and covering of the anterior eye are the mucus-secreting **conjunctiva**. The palpebral conjunctiva are reflected onto the eyeball via **superior and inferior fornices** (L. arched roof). The bulbar conjunctiva cover the front of the eyeball and cornea. Thus, there is a conjunctival sac continuous with the palpebral skin which is closed when the eyes are shut. The red **lacrimal caruncle** at the medial angle of the conjunctival sac consists of modified sebaceous and sweat glands.

The conjunctival sac is continuously bathed in tears secreted by the **lacrimal gland** seated in the lacrimal fossa of the zygomatic process of the temporal bone. Tears flow across the conjunctival sac and are picked up in two small **lacrimal punctae** visible at the medial extremities of both lids. Two short **lacrimal ducts** pass medially above and below the caruncle to enter a **lacrimal sac** situated in a fossa in the bony medial wall of the orbit. This sac is the dilated superior end of the **nasolacrimal duct** which carries the tears to an exit in the inferior meatus of the nose (under the inferior concha). The moisture supplied by tears is essential for the health of the fragile corneal epithelium and cornea proper.

The upper lid is elevated by the **levator palpebrae muscle** (oculomotor nerve). The **orbicularis oculi** muscle closes the eye (facial nerve).

Muscles of facial expression. The subcutaneous striated muscles of the face and scalp are of insignificant bulk and strength, but of very significant function. They open and close facial orifices, and are the chief means of expressing the entire range of emotions. Inability to close the eyes may result in drying of the cornea with ulceration and destruction. A paralyzed face is a distorted or 'dead' face – a very distressing situation for interpersonal relationships if not for somatic health! These muscles are all powered by the facial nerve. Scalp muscles wrinkle the forehead or elevate the eyebrows (see Scalp section above). The muscle that closes the eye is orbicularis oculi, which is thick around the orbital margins but thin in the lids themselves. The elevator of the lid is levator palpebrae, originating from the posterior bony wall of the orbit just above the optic foramen and inserting in the upper lid tarsal plate. Muscles around the nose either flare or close the nostrils. Orbicularis oris encircles and closes the mouth. Depressor anguli oris, levator labia superioris, depressor labia inferioris and risorius are all facial muscles with obvious functions. Zygomaticus major and minor elevate the angle of the mouth whereas depressor anguli depresses it. There are other similar and smaller muscles contributing nuances of facial expression. Buccinator is the muscle of the cheek and is described in the Mouth and Pharynx section.

Facial nerve (cranial nerve VII). This innervates all of the muscles of facial expression (and has other duties as well). It issues from the skull through the stylomastoid foramen and traverses the parotid gland before appearing in the face. Five branches supply the facial muscles and are named for the areas they supply: temporal, zygomatic, buccal, mandibular and cervical, the latter to the platysma of the neck. The facial nerve branches are superficial; they run over the masseter fascia and, if injured, result in varying degrees of facial distortion. Anterior to the masseter muscle is a **buccal fat pad** covering the buccinator (cheek) muscle. The buccal branches of VII (motor) as well as of V3 enter the muscle, the latter emerging from deep to the masseter, and progressing through the cheek to supply sensory fibers to the skin and surface mucosa of the cheeks.

Sensory innervation of the face. The skin of the face is supplied by the trigeminal nerve.

V1 – ophthalmic division – supplies the anterior nose, eyes and forehead mainly through its supraorbital and external nasal branches;

V2 – maxillary division – supplies the lateral nose, cheeks and area extending up into the anterior temples. The nerve branches involved are mainly the infraorbital and zygomaticotemporal;

V3 – mandibular division – supplies the skin of the mandible and posterior temples through its mental, buccal and superficial temporal branches. (The bulk of the ear and the scalp behind the ears are innervated by cervical nerves.)

Arteries of the face. The two major suppliers of the face are the facial and superficial temporal arteries.

Facial artery. A branch of the external carotid, the facial artery originates in the carotid triangle. It crosses the mandible at the anterior margin of the masseter muscle after passing behind the submandibular gland. Angling forwards and upwards, it ends in the area of the medial canthus of the eye. It offers branches to all superficial structures of the face.

Superficial temporal artery. One of two terminal branches of the external carotid (maxillary artery is the other), this is the continuation of the external carotid. Issuing from the substance of the parotid, it crosses the zygomatic process of the temporal bone (where its pulse is readily palpated) and runs up over the temporalis muscle to the skull vertex. A **transverse facial artery** branch courses forwards just inferior to the zygomatic arch to supply the facial structures in this area.

Parotid gland duct. This gland is described in the Neck section. The parotid duct is an important feature of the face. It crosses superficial to masseter, then pierces buccinator to open into the mouth opposite the upper second molar tooth.

Be aware that these vessels and nerves have deep structure responsibilities as well as serving the superficial facial tissues. At the end of this section is a complete plan of the arterial supply of the head. The cranial nerves are also presented in their entirety.

Temporal region.   The area includes two fossae: **temporal** and **infratemporal**.

Temporal fossa. This is the area above the zygomatic arch. The floor (back wall) of the fossa consists of the squama of temporal, greater wing of sphenoid, parietal and frontal bones. The **pterion** is the term used to describe the area where these four bones meet. This is a thin part of the calvaria which is often fractured, with subsequent injury to middle meningeal artery branches and development of epidural hematomas (see Skull section). The fossa contains the **temporalis muscle**, which originates from the bony floor of the fossa and from investing temporalis fascia. Insertion is on the coronoid process of the mandible. It is a powerful jaw-closer and retractor. Temporalis is innervated by deep temporal branches of V3, the trigeminal mandibular division.

Infratemporal fossa. The bony boundaries of this busy place were described in the Skull section. Briefly:

Outside: Ramus of mandible, zygomatic arch
Inside: Lateral pterygoid plate
Front: Posterior aspect of maxilla
Back: Articular tubercle of temporal bone

Above: Continuous with temporal fossa; medially is the greater wing of sphenoid with its foramina ovale and spinosum, plus the undersurface of the temporal squama
Below: Alveolar border of maxilla.

Muscles of mastication. These are temporalis, masseter, and lateral (external) and medial (internal) pterygoids.

Temporalis. Already described (see Temporal fossa), this closes and retracts the jaw.

Masseter. This is not a true inhabitant of the infratemporal fossa, but is described here for the sake of completeness.
Origin: Zygomatic arch
Insertion: Angle and posterior body of the mandible
Nerve: Masseter nerve (V3)
Action: Closes jaw.

Lateral pterygoid.
Origin: Lateral pterygoid plate and undersurface of greater wing of sphenoid
Insertion: Anterior aspect of mandibular condyle and disc of the temporo-mandibular joint
Nerve: Pterygoid branches of V3
Action: Pulls disc and condyle forwards (protracts jaw). Acting singly, the lateral pterygoids move the jaw from side to side (grinding). They open the jaw as permitted by relaxation of masseter, temporalis and the medial pterygoid.

Medial pterygoid.
Origin: Pterygoid fossa (between pterygoid plates)
Insertion: Inner aspect of angle of mandible
Nerve: Pterygoid branches of V3
Action: Jaw-closer in synergy with masseter and temporalis.

Temporomandibular joint (TMJ). Involving the anterior half of the mandibular fossa and articular tubercle of the temporal bone, this joint articulates with the condyle of the mandible. (The posterior half of the mandibular fossa is non-articular.) The joint is supported by the **sphenomandibular ligament** from the sphenoid spine (sentinel of the foramen spinosum) to the lingula of the mandibular ramus and the **stylomandibular ligament** from the styloid process to the mandibular angle. The joint capsule is strengthened by collateral **temporomandibular ligaments**. Dividing the joint into upper and lower halves is a fibrous **articular disc**. The simple hinge motion of the TMJ is obvious. However, when opening the mouth more widely, the condyles and disc glide forward out of the confines of the mandibular fossa to ride under the cartilage-covered articular tubercle with the guidance and confinement of the articular disc. The disc becomes the joint surface for the condyle anterior to the mandibular fossa. With forced opening, the condyles may dislocate entirely and pass forward over the tubercle into the infratemporal fossa. Reduction requires a strong push downwards on the lower molars to stretch the jaw-closers, thus permitting a backwards push to clear the condyles back over the tubercle to the mandibular fossa. The lateral pterygoid attached to the condyle and disc is the prime mover in jaw-opening. Supra- and infrahyoid muscles in the neck are of some help when more force is needed as when opening the jaw against resistance. Noisy clicking TMJs may occur with the disks have worn out for one reason or another (arthritic changes, malocclusion, chronic dislocation, etc.).

Nerves of the infratemporal fossa. The chief neural inhabitants here are branches of the mandibular division (V3) of the trigeminal nerve. In addition, special sensory nerves (taste) and parts of the parasympathetic system can be identified here.

Mandibular nerve (V3) enters the fossa through the foramen ovale in the sphenoid roof of the fossa and immediately emits a shower of small motor branches to the muscles of mastication, tensor tympani and tensor veli palatini. Also from the proximal nerve are the buccal nerve for sensation of the inside and outside surfaces of the cheek, the auriculotemporal nerve for sensation in the temples, and meningeal branches reentering the skull with the middle meningeal artery. Major players, however, are the lingual and inferior alveolar branches of V3.

The **lingual nerve** descends from the foramen ovale on the surface of the medial pterygoid muscle deep to the lateral pterygoid. After escaping from under the latter, it turns forwards to run just under the cover of the mandible body to reach finally the side of the tongue. Here it provides 'ordinary' sensory fibers to the anterior two-thirds of the tongue, and to the surrounding mouth and gums.

Chorda tympani. Shortly after entering the infratemporal fossa and while still behind the lateral pterygoid, the lingual nerve is joined by the chorda tympani, a branch of cranial nerve VII (facial nerve). The chorda tympani contains para-sympathetic preganglionic fibers destined for the submandibular and sublingual salivary glands, and also carries fibers recording taste from the anterior two-thirds of the tongue. As the lingual and its facial nerve branch hitchhikers pass superior to the submandibular ganglion, the parasympathetic preganglionics drop off and enter a small **submandibular ganglion** situated between the lingual nerve and submandibular gland. Here they synapse and, as postganglionics, enter the submandibular and sublingual salivary glands. The taste fibers continue with the lingual nerve without interruption to the taste buds of the anterior two-thirds of the tongue.

The **inferior alveolar nerve** runs adjacent to the lingual in its proximal portion; after escaping the cover of the lateral pterygoid, it diverges posteriorly and enters the mandibular foramen at the base of the mandibular lingula. Traveling in the mandibular canal in the body of the mandible, it supplies nerves to the lower teeth. Emerging from the distal end of the mandibular canal via the mental foramen as the **mental nerve**, it supplies the skin of the chin, and skin and mucous surfaces of the lower lip. (Upper teeth are innervated by branches of the maxillary nerve – V2; see below.)

Mylohyoid nerve. Just before entering the mandibular foramen, the mylohyoid nerve branches off from the inferior alveolar, then travels along the inner aspect of the mandible body to the mylohyoid muscle, which it innervates. Continuing through the muscle, the nerve reaches and innervates the anterior belly of digastric.

Otic ganglion. This tiny ganglion lies alongside the V3 trunk as it leaves the foramen ovale. The ganglion is fed by parasympathetic preganglionics contained in the lesser petrosal branch of the glossopharyngeal nerve (cranial nerve IX). The lesser petrosal passes through the foramen ovale, synapses in the otic ganglion and travels to the parotid gland on the back of the auriculotemporal branch of V3.

Maxillary artery. One of the terminal branches of the external carotid, this runs a short but busy course through the infratemporal fossa. Taking off at right angles to the direction of the external carotid, the maxillary runs through the deep part of the parotid to pass into the infratemporal fossa under the upper mandibular ramus. Continuing forwards, it passes either deep or superficial to the lateral pterygoid to reach the pterygomaxillary fissure. It ascends the fissure and attains the pterygopalatine fossa, then disappears through the sphenopalatine foramen (as the sphenopalatine artery) to the nasal cavity. A multitude of named branches issues from the maxillary. The first important branch is the middle meningeal artery, given off behind the mandible, and ascending to and through the foramen spinosum in the roof of the infratemporal fossa.

The maxillary artery crosses superficial to the inferior alveolar and lingual nerves, sending an inferior alveolar branch to run with the nerve of the same name through the mandibular canal. In the pterygopalatine fossa, posterior superior alveolar branches and the infraorbital artery appear. The latter accompanies the infraorbital nerve in the infraorbital groove, canal and foramen. En route, it gives off anterior superior alveolar arteries. In addition, while in the pterygopalatine fossa, a greater palatine artery arises and accompanies the greater palatine nerve to the palate. Numerous other maxillary artery branches supply blood to, for example, the infratemporal muscle and bone, auditory meatus and middle ear, cheek, pterygoid canal, eustachian tube and pharynx.

Pterygoid venous plexus. A considerable mass of veins surrounds the maxillary artery and lateral pterygoid muscle. The importance of this venous plexus lies in its connections with intracranial veins as well as the jugular system. Branches travel with the mandibular nerve through the foramen ovale to the cavernous sinus, with the middle meningeal artery to the cerebral meninges, and via the buccal veins accompanying the buccal artery to connections with the facial and ophthalmic veins. Infection in the face can spread via infected venous blood to the brain with dire consequences.

Pterygopalatine fossa. This area was described in the Face section of the description of the Skull. We will now put it to work! Remember that this space connects with the infratemporal fossa via the pterygomaxillary fissure, with the orbit via the infraorbital fissure (which is like a large skylight in the pterygopalatine fossa roof), with the nasal cavity via the sphenopalatine foramen, and with the middle cranial fossa via the foramen rotundum in the fossal back wall. The maxillary nerve – V2 – entering the fossa through the rotundum is described first, followed by the autonomics in the fossa and finally the fossal blood vessels.

Maxillary nerve. After entering the fossa through the foramen in the base of the greater wing of sphenoid, this totally sensory nerve passes across the top of the fossa to enter the orbit via the infraorbital fissure. Here it becomes the infraorbital nerve and runs in the familiar infraorbital fissure-groove-tunnel-foramen path to reach the face. Within the fossa, the nerve gives off its **zygomatic branch**, which divides into **zygomaticofacial** and **zygomaticotemporal** branches for the upper cheek, side of the nose and anterior temple. The nerves exit via small similarly named foramina in the zygoma. Next in the fossa come the **pterygopalatine nerves**, which have fibers bound for the nose and palate, and are described below. En route in the pterygopalatine foramen, they pass without synapse through a parasympathetic ganglion, the **pterygopalatine ganglion** (see below). Finally, **anterior and posterior superior alveolar nerves** are seen branching from the maxillary trunk and infraorbital nerve to supply the upper teeth.

Pterygopalatine ganglion. This parasympathetic ganglion is 5 mm in diameter and 'suspended' in the pterygopalatine fossa by the pterygopalatine nerves. The ganglion is 'fed' by the **nerve of the pterygoid canal**. The latter structure was described previously as consisting of preganglionic facial nerve fibers (greater pretrosal nerve) issuing from the facial hiatus in the anterior face of the petrous pyramid near its apex and sympathetic postganglionic fibers carried on the back of the internal carotid artery (deep petrosal nerve). (The deep petrosal joins the greater petrosal just outside the facial hiatus.) The two petrosals form the nerve of the pterygoid canal and enter the canal at the posterior aspect of the base of the sphenoid body. The canal passes forwards in the sphenoid body and under the sphenoid sinus to open in the posterior wall of the pterygopalatine fossa. The nerve enters the **pterygopalatine ganglion** where the facial parasympathetic preganglionic portion synapses. The postganglionics of both autonomic divisions then run to the lacrimal gland and palatonasal mucosa. To reach the lacrimal gland, these autonomic fibers first join the maxillary trunk, then its zygomatic branch, and finally the lacrimal branch of V1 to attain their lacrimal destination. The autonomic road to the nasal cavity and palate is easier as it uses branches of the pterygopalatine nerves.

Maxillary artery. The maxillary climbs the pterygomaxillary fissure to reach and enter the pterygopalatine fossa. The artery in the fossa emits branches which accompany the branches of the maxillary nerve – **superior alveolar, infraorbital** and **greater palatine**. In addition, two tiny vessels, the **artery of the pterygoid canal** and a **pharyngeal branch**, are given off in the fossa. After all this, the artery passes through the sphenopalatine foramen in the medial wall of the fossa to become the **sphenopalatine artery**, and supplies blood to the nose and palate. Maxillary veins accompany these artery branches.

Nasal cavity. We will first describe the bones and cartilage, then the sinuses, and finally the nerves and vessels.

Medial wall: Nasal septum. The nasal septum is composed of bone posteriorly and cartilage anteriorly. The bones are the **perpendicular plate of the ethmoid** above and the **vomer** below. Both articulate with the sphenoid posteriorly whereas anteriorly they meet septal cartilage. Inferiorly the vomer meets the midline crests of the maxilla and palatine bone. Anterosuperiorly, the ethmoid meets the nasal bones. The septal cartilage continues forwards to split the nares.

Roof. This is made up of **nasal bones** in front, the **cribriform plate** of the ethmoid in the middle and the sphenoid body behind. The cribriform plate is marked by multiple openings for the olfactory nerves.

Floor. The anterior four-fifths is the **palatine process of the maxilla**; the posterior one-fifth is the **horizontal process of the palatine.** In the anterior end of the maxillary portion is found the **incisive canal,** which transmits the V2 nasopalatine nerve, supplying sensory innervation to the anterior palate and neighboring gums (see below).

Lateral wall. The maxilla lies in front, and in the midportion are the maxilla, **superior and middle conchae of the ethmoid** and **inferior concha.** Behind are the vertical portion of the palatine and medial pterygoid plate of the sphenoid. The inferior concha is a bone of its own whereas the middle and superior conchae are processes of the ethmoid. The superior concha is the smallest, the inferior the largest of the three. The three are more or less horizontally disposed and overhang three channels or meatus. The largest, the **inferior meatus,** shows the opening of the **nasolacrimal duct** anteriorly (from the lacus lacrimalis at the inner canthus of the eye) for relieving tears produced by the lacrimal gland. (Do you ever blow your nose after seeing a sad movie?) The smaller **middle meatus** demonstrates in its midportion a crescent-shaped fissure, the **hiatus semilunaris.** Into the anterior extremity of the hiatus empties the frontonasal duct, which drains the frontal sinus. Further posteriorly in the hiatus is the opening of the maxillary sinus. Middle and anterior ethmoid cells also empty into the middle hiatus. The smallest of the three, the **superior meatus,** receives effluent from the posterior ethmoid sinuses and contains the nasal opening of the sphenopalatine foramen.
Anteriorly and posteriorly, the nasal cavity is open through the nares and the choanae.

Nasal sinuses. These paired, but asymmetrical, paranasal air cells serve to help filter, warm and humidify inhaled air, lend resonance to the voice and lighten the structure of the face. They are connected to the nasal cavity as detailed above and include the frontal, ethmoid, sphenoid and maxillary sinuses.

Frontal sinuses. These are totally asymmetrical and have a wide range of size (from a few millimeters to more than 5 cm in height). They lie in the frontal bones behind the superciliary arches. The frontonasal ducts connect the sinuses with the anterior portion of the hiatus semilunaris of the middle meatus.

Ethmoid sinuses. These are numerous small air cells situated between the upper nasal cavity and orbit. Divided into anterior, middle and posterior groups, the cells are encased in thin bone. The lateral masses of the ethmoid are filled with these small cells so as to confer a labyrinthine nature to the area. The thin lateral coverings of the ethmoid cells form a large part of the paper-thin medial orbital wall. The anterior and middle groups drain into the middle meatus whereas the posterior group drains into the superior meatus.

Sphenoid sinuses. These larger (average size 2 x 2 x 2 cm) air cells reside in the body of the sphenoid. They open into a recess just superior to the superior meatus. The **pterygoid canals** pass under the sphenoids, often producing visible ridges in the sinus floors.

Maxillary sinuses. These are the largest sinuses of the group. They occupy the body of the maxilla on each side. The sinus roofs lie immediately under the orbit floors. The sinus floors are the maxillary alveolar processes and have indentations made by the roots of the upper teeth. Each maxillary sinus has a capacity of approximately 15 ml. The maxillary cells drain into the middle meatus.

The mucous membrane of the nasal cavity is thick, juicy and vascular. Swelling due to nasal inflammation increases the thickness of the membrane, and often occludes the sinus openings and the nasal airway itself. Recognizing the close relationship of the ethmoids to the orbital medial wall, and the maxillaries to the oribital floor and to the teeth, it should not be surprising that aching eyes, teeth and cheeks are common accompaniments of disorders such as the common cold. All cold medicines contain vasoconstrictor mucosa-shrinking agents.

Fractures of the zygoma and orbital floor are frequent and often involve the maxillary sinus. Bone fragments, eye muscle, orbital fat and blood may be forced into the sinus. The eye may be depressed and its movements inhibited in such injuries. Abscessed teeth may drain into the sinus through the sinus floor.

Nerves of the nasal cavity.

Somatic trigeminal (V2) branches. We left the maxillary nerve running across the top of the pterygopalatine fossa and mentioned two **pterygopalatine** branches suspending the pterygopalatine ganglion there. The two branches unite at the ganglion, then break up into multiple branches to supply the nasal cavity.

Greater palatine nerve. This branch descends from the pterygopalatine fossa to the hard palate through a tunnel at the junction of the maxilla and palatine bones called the **greater palatine canal**. The tunnel ends at the **greater palatine foramen** in the palatine portion of the hard palate just medial to the third molar tooth. The greater palatine nerve supplies sensory fibers to the palate and gums, and adjacent soft palate. En route through the canal, **posterior inferior nasal branches** are given off which supply the inferior conchal area.

Lesser palatine nerves. These pursue a similar course as the greater palatines. They innervate the posterior soft and hard palate, and the tonsils.

Posterior superior palatine nerves. These branches pass through the sphenopalatine foramen to innervate the upper posterior nasal cavity and posterior nasal septum. A long offshoot – the **nasopalatine nerve** – passes across the upper nasal cavity to reach the upper nasal septum. From there, it proceeds forwards and downwards, innervating the septum along the way, to finally reach and pass through the incisive canal. Beyond the canal, it innervates the front of the gums and palate.

The remainder of the nasal septum – the anterior portion – is innervated by the **anterior ethmoid branch** of the **nasociliary nerve** (V1; see Eye orbit section). Ethmoid branches of this nerve also innervate the sphenoid, ethmoid and frontal sinuses.

Olfactory nerve (cranial nerve I). Olfactory nerve endings congregate in the roof of the nasal cavity, over the superior concha and opposing septum. The nerves pass through the ethmoid cribriform plate to reach the intracranial olfactory bulb.

Autonomics of the nasal cavity. The multitude of mucous and serous glands of the nasal and palatal mucosa require autonomic control. Sympathetic postganglionics and parasympathetic preganglionics reach the pterygopalatine fossa as the **nerve of the pterygoid canal**. Parasympathetics synapse here in the pterygopalatine ganglion whereas sympathetics pass on through. Exiting postganglionic fibers travel with blood vessels and trigeminal branches to reach the glands.

The lacrimal gland is also a responsibility of the nerve of the pterygoid canal and the pterygopalatine ganglion. Postganglionic autonomic fibers reach the lacrimal gland by traveling with the maxillary nerve to its zygomatic branch. Together with this nerve, they are carried to the lacrimal nerve of V1, which then carries them to their destination.

Arteries and veins of the nasal cavity. Only the sphenopalatine artery – the terminal branch of the maxillary artery – is left. The maxillary changes its name as it passes through the sphenopalatine foramen on leaving the pterygopalatine fossa. (The descending palatine artery, which accompanies the greater palatine nerve in the greater palatine canal, is given off in the pterygopalatine fossa before the main vessel passes through the sphenopalatine foramen.) Branches of the descending palatine and sphenopalatine arteries accompany nerves supplying the lateral wall of the cavity, septum and palate. Ophthalmic artery **ethmoid arterial branches** accompany the V1 nerves of the same name to nourish the ethmoid, frontal and sphenoid sinuses, and anterior septum.

In general, veins run with the arteries.

Mouth and oral cavity.
Everybody knows where this is! Regard it as the beginning of a long muscular mucosa-lined tube extending from the mouth to the anus. We start by reviewing the major nerves and vessels supplying the area.

Lingual nerve (V3). This passes to the side of the tongue sheltered by the body of the mandible. It supplies sensory fibers to the anterior two-thirds of the tongue, and surrounding gums and mucosa.

Mylohyoid nerve. This nerve branches from the V3 inferior alveolar nerve just before the mandibular foramen. It passes down under cover of the body of the mandible to innervate the mylohyoid and anterior digastric muscles.

Buccal nerve. This is a branch from the trunk of V3. It passes through the lateral pterygoid muscle to supply sensory fibers to both the inside and outside surfaces of the cheek. (The buccal branch of cranial nerve VII innervates the buccinator muscle.)

Chorda tympani. A branch of cranial nerve VII (facial nerve), this carries taste from the anterior two-thirds of the tongue, and preganglionics for the submandibular ganglion.

Postganglionics from the ganglion power the submandibular and sublingual salivary glands.

Hypoglossal nerve. Starting at the opening of the hypoglossal canal deep to the internal carotid and internal jugular, the nerve runs forwards in the neck and becomes superficial near the angle of the mandible, where it appears from under the posterior belly of digastric. It arcs down over the external carotid and lingual arteries, then loops back up to disappear under digastric and stylohyoid. From there, it runs forwards, slipping under the free border of mylohyoid to finally reach the tongue and its muscles.

Glossopharyngeal nerve (cranial nerve IX). This nerve supplies taste and general sense for the posterior one-third of the tongue (described in detail with the pharynx).

Facial artery. This external carotid artery branch appears at the anterior margin of the masseter as it crosses the mandible. It then ascends across the cheek to the region of the inner angle of the eye to distribute superficial and deep branches to structures of the face.

Lingual artery. Another branch of the external carotid, this passes over the hyoid bone, then forwards under the digastric and stylohyoid muscles to enter the tongue, running forwards all the way to its tip. Its branches supply tongue muscle and mucosa, the sublingual glands and, posteriorly, the faucial pillars and tonsils.

Cheeks. These are the lateral confines of the mouth. Between the skin mucosa and skin lies the buccinator muscle with its buccal branch innervation from the facial nerve. The cheeks serve to keep food between the teeth during mastication (described in detail with chewing and swallowing).

Floor of the mouth. This muscular floor consists of the **mylohyoid** and **geniohyoid** muscles. The former originates from the mylohyoid line of the mandible along the arc of the mandible from third molar to third molar. Insertion is at the hyoid bone and its nerve is the mylohyoid branch of V3. Its action is to raise the hyoid bone and tongue. The geniohyoid consists of a pencil-thin pair of muscles extending from the symphysis menti to the hyoid. Its nerve is C1 (via the hypoglossal) and its action is to draw the hyoid and tongue forwards. (Again, more about these muscles with chewing and swallowing.)

Roof. This is the hard palate.

Anterior extremity. This is the lips.

Teeth. The full adult dentition consists of 32 teeth: 4 incisors; 2 canines; 4 premolars; and 6 molars per each upper and lower jaw.

Tongue. The diffusely and homogeneously papillated dorsal surface ends at the V-shaped line of large **vallate papillae** demarcating the posterior boundary of the anterior two-thirds of the tongue. At the apex of the V is a dimple, the **foramen cecum**, site of the thyroid anlage. The posterior third is smoother and studded with larger projections of lingual lymphoid tonsillar material. Holding the tongue down to the oral cavity floor is a thin midline vertical fold of mucosa called the **frenulum**.

Muscles of the tongue. These comprise both extrinsic and intrinsic muscles. The extrinsic muscles are mainly genioglossus and hyoglossus. **Genioglossus** originates from the mental symphysis and fans up into the tongue. Its posterior fibers protrude, and anterior fibers retract, the tongue. **Hyoglossus** passes from the hyoid to the sides of the tongue, where it controls the action. **Styloglossus**, extending from the styloid process to the tongue, draws the tongue up and back.

Intrinsic muscles fine-tune tongue activity.

Acting as a group, these muscles provide the wide range of movements exhibited by the tongue.

All tongue muscles are innervated by the hypoglossal nerve.

Salivary glands. The orifice of the **parotid duct** is in the cheek opposite the second upper molar.

The orifice of the **submandibular duct** (Wharton's duct) is on the side of the base of the frenulum atop a small papilla.

The **sublingual** salivary gland is 1–2 cm long and extends under the mucosa of the junction of the floor of the mouth and base of the tongue. The gland rests on the mylohyoid muscle and drains through multiple small openings along its course. Secretion of the parotid duct is controlled by parasympathetics from the glossopharyngeal by way of the otic ganglion. The submandibular and sublingual are controlled by chorda tympani parasympathetic fibers from the facial nerve via the submandibular ganglion.

Nerve and artery supply.

The **lingual nerve** enters the oral area lateral to the submandibular duct, then hooks around it to enter the tongue. The **lingual artery** enters the base of the tongue at the level of the hyoid to supply the deep tongue muscles, then ascends to supply the more superficial tongue structures.

The **glossopharyngeal nerve** issues from the jugular foramen. Its tongue portion follows and innervates the stylopharyngeus muscle, then turns across the muscle to enter and pierce hyoglossus to reach the posterior third of the tongue. The sensory ganglia of the glossopharyngeal lie in the jugular foramen.

The nerve supply of the oral cavity may be summarized as follows. Somatic sensory fibers are supplied by the lingual nerve to the mucosa and anterior two-thirds of the tongue, by the buccal branch of V3 to the cheeks and gums, and by the palatines and alveolars of V2. The glossopharyngeal serves the posterior third of the tongue. Taste is supplied by the facial nerve chorda tympani for the anterior two-thirds and by glossopharyngeal for the posterior third of the tongue. Motor activity of the tongue is governed by the hypoglossal. The facial nerve innervates orbicularis oris and buccinator. Salivary secretory activity is ordered by the chorda tympani for the submandibular and sublingual glands, and by the glossopharyngeal for the parotid. (The pharyngeal plexus – vagus – raises the soft palate.) Examination of the oral cavity can reveal much information regarding the integrity of cranial nerves V, VII, IX, X and XII!

Fauces (L. throat).   This is the space between the oral cavity and pharynx. Above is the soft palate and below is the posterior dorsum of the tongue. On either side are the faucial arches.

Soft palate and **faucial arches**.   The flexible soft palate is a movable curtain hanging from an anterior attachment to the hard palate. It is supported by a sheet of fibrous tissue, the **palatine aponeurosis**, which is attached anteriorly to the hard palate. The inferior free border is marked in its midline by the fleshy uvula (L. little grape). The remainder of the free margin arches laterally to two mucosal folds, the **palatoglossal** and **palatopharyngeal arches**, or **faucial pillars**. The soft palate and arches contain muscles: levator and tensor veli palatini (*velum* = L. veil or curtain) of the palate; palatoglossus and palatopharyngeus in the arches. **Levator veli palatini** originates at the undersurface of the apex of the petrous pyramid just anterior to the carotid canal and medial lamina of the cartilaginous eustachian tube. The muscle passes down around the choanal border to enter the soft palate and insert in the palatine aponeurosis. Its nerve supply comes from the vagus (see below in the pharyngeal plexus section). **Tensor veli palatini** arises from the base of the medial pterygoid plate and spine of sphenoid, and intervening anterior lip of the eustachian tube cartilage. The muscle descends to the inferior extremity of the medial pterygoid plate, where it winds around the pterygoid hamulus and makes a right-angle bend to enter the soft palate from the side. Its nerve supply is trigeminal. The tensor is lateral and anterior to the levator. The **palatoglossus** muscle runs from the soft palate to the side of the tongue. **Palatopharyngeus** also originates in the soft palate, fills the posterior palatal arch and continues with stylopharyngeus down to the thyroid cartilage and side of the laryngopharynx. Both arch muscles are innervated by the pharyngeal plexus. Actions of these palatal and arch muscles stiffen and/or raise the soft palate and constrict the faucial opening (isthmus). Thus, the oropharyngeal or nasopharyngeal orifice may be effectively closed off.

Tonsil.   Situated between the faucial arches is the **palatine tonsil**. The tonsil is part of a ring of immunologically active lymphoid tissue guarding the entrances of the air and food passages. The remainder of the ring includes the already mentioned lingual tonsillar tissue over the posterior third of the tongue, and the pharyngeal tonsils (adenoids) clustered about the auditory tube area of the nasopharynx. The connection between tonsils and poliomyelitis is well known. The almost ritual T & A (tonsillectomy and adenoidectomy) of years gone by has been replaced by efforts to save and preserve tonsillar tissue wherever possible. The blood supply of the palatine tonsil includes branches from all major arteries of the area: lingual (posterior tongue branches); facial (ascending palatine and tonsillar branches); maxillary (branches of the descending palatine); and external carotid itself (ascending pharyngeal).

Pharynx.   This is the funnel-shaped muscular tube behind the nose, mouth and larynx. Its open end consists of the openings of the nasal and pharyngeal passages; its borders are attached to the walls of these passages and base of the skull. The funnel narrows down to its outlet, the esophagus. Described first are the muscles of the tube, followed by details of the mucosal surfaces of the nasal, oral and laryngeal areas of the tube.

Pharyngeal muscles. These begin with the buccinator muscle of the cheeks. The remainder consists of three wraparound constrictor muscles and several vertically disposed elevator muscles.

Superior constrictor. Anteriorly this muscle joins buccinator in a fibrous **pterygomandibular raphe** which extends from the pterygoid hamulus to the mandible. The muscle fibers curve backwards to meet in a posterior midline raphe. In the posterior midline, the upper fibers arch upwards and attach to the **pharyngeal tubercle** of the occipital bone. A **pharyngobasilar fascia** fills the muscle-free interval between the muscle attachments at the tubercle and pterygoid hamulus.

Middle constrictor. This muscle originates from the entire length of the greater cornu of the hyoid bone, and the lesser cornu and stylohyoid ligament which inserts on the lesser cornu. Muscle fibers curve around to the posterior midline raphe, overlapping the lowermost fibers of the superior constrictor.

Inferior constrictor. The lowest constrictor arises from a diagonal ridge (oblique line) on the side of the thyroid cartilage and side of the cricoid. (The other side of the ridge serves as the insertion of the sternothyroid muscle.) Here also the muscle fibers arch around to the posterior midline raphe, overlapping the lowermost fibers of the middle constrictor. The lower fibers of the inferior constrictor are transverse in contrast to the upward-tending upper fibers. The transverse fibers blend with the muscle of the esophagus. This **cricopharyngeus** portion serves as an upper esophageal sphincter.

The vertically disposed pharyngeal muscles are relatively small thin muscles.

Stylopharyngeus. This slender muscle originates at the styloid process and proceeds downwards between the superior and middle constrictors to the posterior border of the thyroid cartilage and neighboring inferior constrictor muscle.

Palatopharyngeus. Already mentioned as the muscle in the posterior faucial pillar, this muscle joins stylopharyngeus below the superior constrictor. The two muscles insert together.

Salpingopharyngeus. Arising from the auditory tube, this muscle passes downwards to blend with palatopharyngeus.

There we have the muscular pharynx. Above, between and below each constrictor are clefts for the passage of important structures to the interior. Above the superior constrictor and passing through the pharyngobasilar fascia are the levator veli palatini and auditory tube. Between the superior middle constrictors are the stylopharyngeus muscle, glossopharyngeal nerve and stylohyoid ligament. Between the middle and inferior constrictors are the internal branch of the superior laryngeal nerve and accompanying superior laryngeal artery. Beneath the inferior constrictor passes the recurrent laryngeal nerve.

Pharyngeal plexus. Cranial nerves IX and X, and the bulbar (or accessory) part of cranial XI all supply the pharyngeal muscles. They emit branches which gather on the surface of the middle constrictor to form a pharyngeal nerve plexus. From the plexus, branches of uncertain 'pedigree' issue to the pharyngeal musculature. Although it is certain that the glossopharyngeal nerve innervates stylopharyngeus and that the trigeminal serves the tensor veli palatini, all the other muscles are innervated by nerves of uncertain origin and may sensibly be described as being innervated by the pharyngeal plexus. Furthermore, sensory areas are indistinct. It is customary to assign sensation in the oropharynx to the glossopharyngeal and those in the laryngopharynx to the vagus.

Retropharyngeal space. Posterior to the muscular pharyngeal funnel is a loose areolar tissue space which permits movement of the structure. The clinical significance of the space is that it provides the possibility of easy spread of bacteria up to the base of the skull and down into the thorax. Perforation of the posterior pharynx or esophagus can produce devastating sepsis.

The pharynx is divided into nasopharynx, oropharynx and laryngopharynx.

Nasopharynx. This area lies posterior to the nose and superior to the soft palate. On its lateral wall is the orifice of the auditory (eustachian) tube, projecting into the pharynx by virtue of its surrounding tubal cartilage. On the posterior wall is a protrusion of lymphoid tissue – the pharyngeal tonsil. Enlarged in children, this tissue may block the auditory tube, resulting in middle ear infection. Enlarged pharyngeal tonsils are called adenoids (Gk. in the form of a gland). The tubal cartilage offers origin to the salpingopharyngeus muscle (*salpinx* = Gk. tube).

Oropharynx. Extending from the soft palate to hyoid bone, this is open anteriorly to the mouth through the fauces. The palatine tonsil and faucial arches are considered parts of the oropharynx.

Laryngopharynx. Extending from the hyoid bone to the lower border of the cricoid cartilage, the entry to and structure of the larynx is contained in the laryngopharynx (see above section describing the larynx). Alongside the larynx are the **piriform sinuses**, longitudinal recesses through which swallowed material reaches the esophagus. The thyroid cartilage and thyrohyoid membrane are the lateral boundaries of the piriform sinuses.

Chewing and swallowing. This complex action is in part voluntary and in part reflex. The voluntary part is simple enough: the tongue is used to mobilize a bolus of food and push it between the teeth; buccinator is used to hold the bolus between the teeth; and tensor veli palatini and palatoglossus tighten the soft palate and close the fauces to keep the food in the mouth. When chewed and ready to swallow, the food is pushed to the back of the oral cavity by the tongue. After relaxing the fauces to open the faucial isthmus, the food is pushed into the oropharynx.

This triggers the reflex part. The nasal airway and entrance to the larynx must both be closed, and the pharynx elevated and widened to accept the bolus of food. Pharyngeal constrictors then generate a peristaltic wave to propel food into the esophagus. Be aware that many of the little muscles of the head and neck previously described are involved in this complex procedure of deglutition. The suprahyoid muscles (digastric, stylohyoid, geniohyoid and mylohyoid) manage the hyoid bone and position of the base of the tongue. Vertical pharyngeal muscles elevate and help open the pharynx (stylopharyngeus, palatopharyngeus and salpingopharyngeus). Palatoglossus and palatopharyngeus close the fauces. Tensor and levator veli palatini stiffen and move the soft palate. Intrinsic laryngeal muscles and cartilages with the help of the thyrohyoid muscle control the epiglottis and glottis to prevent entry of food into the tracheal airway. Sternohyoid, sternothyroid, thyrohyoid and omohyoid steady the hyoid bone as necessary.

The act of swallowing may be described as follows. After mastication, the food is passed to the back of the mouth by the tongue. The hyoid bone, base of the tongue and pharynx are raised and pulled forward, thus opening the inlet of the pharynx. The soft palate is plastered up against the posterior pharyngeal wall, closing off the nasal cavity. After the food passes into the oropharynx, the fauces are closed by the muscle of the faucial pillars. After elevation and fixation of the hyoid by the suprahyoid neck muscles (digastric, stylohyoid, mylohyoid and geniohyoid), the larynx is elevated and closed by the actions of the thyrohyoid, stylohyoid and palatopharyngeus, the intrinsic muscles of the larynx, and the epiglottic 'lid' with its aryepiglottic folds. Food is propelled alongside the closed larynx through the piriform sinuses by peristaltic action of the constrictors, and the continued elevation (pulling up) of the pharynx over the food by palatopharyngeus and its elevator relations.

Eye.   The bony orbit is described in the Orbit section of the Skull. The external features (eyelids, conjunctivae, lacrimal apparatus) are described in the Face section. Note that the eye is an extension of the brain. The dura and pia are represented in the eye as the sclera and choroid. The 'optic nerve' is a brain tract; the 'true' optic nerve consists of the neural cells and processes of the retina.

The eye is described here in the following order: the covering layers; refracting media; and muscles, nerves and vessels.

Coverings of the eye.   These include the sclera, vascular coat and retinal layers.

Sclera.   The sclera is the tough fibrous outer cover of the eyeball. Embryologically, this layer is an extension of the dura. It is visible as the white of the eye. At the region of the corneoscleral junction, a tiny 'dural sinus', the **sinus venosus sclerae** (more popularly known as the **canal of Schlemm**), runs around the outer margin of the cornea. It is connected to scleral veins and the venous drainage system of the brain. Aqueous humor (see below) is picked up here in the same way that cerebrospinal fluid is picked up in the brain dural sinuses. Obstruction of the canal causes a back-up of aqueous humor and an increased pressure within the eye (*cf* hydrocephalus of CSF circulatory obstruction). The resulting condition in the eye is the common **glaucoma**, which must be treated medically and/or surgically to prevent destruction of the fragile retina.

Cornea.   This anterior extremity of the sclera is formed by the junction of conjunctiva and sclera. Its transparent nature is obvious. It has lost its blood vessels and sucks up nourishment from capillaries at its perimeter. Ideally, corneal curvature is steady and smooth, although 'wrinkles' and irregularities are occur frequently and require optometric correction for clear vision. The delicate cornea must be kept moist by tears and lids as drying, ulceration and scarring may destroy corneal transparency.

Vascular coat [uveal tract (*uva* = L. grape)]. This layer consists of the choroid, ciliary body and iris.

Choroid. This thin, vascular, pigmented brownish-black layer is embryologically derived from the pia mater. Its vascularity keeps the retina in good health. It extends over and is attached to the pigmented layer of the retina. The pigment of both the choroid and retina serves to eliminate reflection of entering light rays. The light-sensitive retinal nervous tissue ends near the front portion of the eyeball where light cannot strike. Here the chorioretinal layer ends in a jagged circular **ora serrata**, the visible mark of the junction between the chorioretina and ciliary body.

Ciliary body. This structure is made up of choroidal elements, smooth muscle and 'ciliary processes'. It extends from the ora serrata to the lateral border of the iris at the sclerocorneal junction. The ciliary body contains **ciliary processes,** which produce the aqueous humor, and a circumferential smooth muscle – **ciliaris** – which is attached to the lens by suspensory ligaments (grandly termed the **zonule of Zinn**). Left to itself, the lens rounds up into a fat round ball. Contraction of ciliaris constricts the circle of attachment of the suspensory ligaments and allows the lens to 'fatten' whereas pull on the suspensory ligaments by dilation of the ciliary muscle circle flattens the lens. Clearly, the chief dynamic mechanism of visual accommodation – focusing – is the responsibility of the ciliaris muscle of the ciliary body. Ciliaris nerve supply is described below.

Ciliary processes. These short longitudinal bodies are arranged circumferentially in the ciliary body just behind the iris. They secrete aqueous humor in a manner analogous to the secretion of cerebrospinal fluid by the choroid plexuses of the brain.

Iris. This thin pigmented disc-like extension of the choroid is suspended between the cornea and lens in the aqueous humor. Its central opening is the **pupil** of the eye. The iris is invested with smooth muscle which constricts or dilates the pupil. Its position divides the aqueous humor-filled space between cornea and lens into **anterior** and **posterior chambers,** which communicate through the pupil. The nerve supply of ciliaris and the iris smooth muscle is from the autonomic system. Preganglionic parasympathetic fibers are carried from the midbrain in the oculomotor (III) nerve. They enter the ciliary ganglion in the orbit behind the eye and synapse before entering the eye as postganglionic **short ciliary nerves.** Sympathetic postganglionics from the superior cervical ganglion approach the eye on the backs of the internal carotid and ophthalmic arteries. They reach the eye as **long ciliary nerves** without further synapse. The pigment of the iris restricts entry of light except through the pupil.

Neural coat: Retina. The retina is the innermost tunic of the eye. It is attached externally to the choroid and extends anteriorly to the ora serrata. Its medial internal relation is the vitreous body. In the posterior retina in the visual axis of the eye is a small pit, the **fovea** (L. pit). The fovea is surrounded by a yellowish area, the **macula lutea** (L. yellow spot). In the fovea–macula area, vision is at its sharpest and most colorful. A few millimeters nasal to the macula is the grayish-white **optic disc**, the site of exit of the optic nerve fibers, the optic nerve head, the visual 'blind spot'. In the center of the disc, the visible **central artery of the retina** (from the ophthalmic) enters the eye and arborizes in the vascular layer to nourish the retina. Failure of this vessel results in death of the retina and, thus, blindness of the involved eye. The retina is populated by light- and color-sensitive receptors, the **rods** and **cones**, located in the deepest layer of the retina. Four further layers of neural cells, lying more superficial to the layer of the receptors, 'doctor' retinal reception until a final 'stratum opticum' is reached. Cells in this retinal layer furnish the nerve fibers that stream to the optic disc and optic nerve.

Refraction structures. These include the cornea, aqueous humor, lens and vitreous body. The cornea, the watery aqueous humor and the lens are described above.

Vitreous body. This semigelatinous transparent body fills the retinal portion of the eyeball and is attached to the retina. It contains no blood vessels, but receives nourishment from the retinal and ciliary body vessels. The vitreous body holds the retina in place. Loss of vitreous subsequent to trauma or eye surgery, for example, is followed by detachment of the retina, a situation which must be promptly corrected or the detached retina will atrophy. (Choroidal exudates or hemorrhage and tumor are other causes of detachment.)

Extraocular muscles. Be aware that, when the eyes are looking straight ahead at rest, they are at an angle to the axis of the orbit. Note also that the eye can move around three axes: horizontal (elevate/depress); vertical (abduct/adduct); and anterior–posterior (intorsion/extorsion). Thus, unless the axis of the eye is in a direct line with muscle pull, the muscle can produce compound eye movements: one muscle may roll the eye inward in adduction and also elevate it. To effect a straight up or down movement, two muscles are necessary: one to effect elevation or depression; the other to correct intorsion or extorsion, and effect adduction.

This complicated situation can be bypassed when examining eye muscle function by first turning the eye so that its axis is in line with the pull of the muscle to be tested. Thus, contraction of the muscle will produce movement only in the direction of its pull. By abducting the right eye (lateral rectus muscle and abducens nerve), the superior and inferior rectus muscles are placed in line with/parallel to the visual axis. In this abducted position, contraction of the superior rectus produces only elevation of the eye; no rotatory motion is imparted. Similarly, depression requires only contraction of the inferior rectus muscle. This is followed by adduction of the right eye (medial rectus). In this position, tendons of the superior and inferior obliques are in line with the visual axis. Pull by the superior oblique (trochlear nerve) depresses, and pull by the inferior oblique elevates, the adducted eye; no other axial movement occurs and no other muscle is required.

The oculomotor nerve is tested in movements of the medial, superior and inferior rectus muscles, and inferior oblique (and by the presence or absence of parasympathetic pupillary constriction).

Levator palpebrae
Origin: Lesser wing of sphenoid just above the optic foramen
Insertion: Superior tarsus
Nerve: Oculomotor (III)
Action: Elevates upper lid.

The four rectus muscles originate from a common annular tendon surrounding the optic foramen.

Superior rectus
Origin: Common annulus
Insertion: Superior eyeball
Nerve: Oculomotor (III)
Action: Elevates, intorts, adducts

Inferior rectus
Origin: Common annulus
Insertion: Inferior eyeball
Nerve: Oculomotor (III)
Action: Depresses, extorts, adducts

Medial rectus
Origin: Common annulus
Insertion: Medial eyeball
Nerve: Oculomotor (III)
Action: Medially rotates (adducts)

Lateral rectus
Origin: Common annulus
Insertion: Lateral eyeball
Nerve: Abducens (VI)
Action: Laterally rotates (abducts)

Superior oblique
Origin: Sphenoid medial to annulus
Insertion: Muscle passes forwards along the medial orbit wall to reach a fibrous ring or trochlea (Gk. pulley) at the upper inner aspect through which it passes, then doubles back over the eye to insert in the lateral eyeball just posterior to the horizontal equator of the eye
Nerve: Trochlear (IV)
Action: Depresses, intorts, abducts

Inferior oblique
Origin: Anterior orbital floor lateral to nasolacrimal canal
Insertion: Under the eye to the lateral eyeball posterior to the horizontal equator of the eye
Nerve: Oculomotor (III)
Action: Elevates, extorts, abducts.

It is evident that looking straight up with the eyes straight ahead requires both the superior rectus and inferior oblique muscles. Both elevate the eyeball, but the extortion/abduction of the inferior oblique opposes and negates the intortion/ adduction of the superior rectus.

Autonomic nerves of the eye. The smooth muscle in the ciliary body and iris require autonomic innervation. Parasympathetic preganglionics originate in the midbrain Edinger–Westphal nucleus and are carried in the oculomotor nerve to the orbit. Here they synapse in the small **ciliary ganglion** lying adjacent to the optic nerve. Postganglionics emanating from the ganglion proceed to the eye as the **short ciliary nerves.** Sympathetic preganglioncs originate in the lateral horn cells of T1. Fibers proceed via a white ramus communicans to the sympathetic trunk, which they ascend to reach the superior cervical ganglion. After synapse there, postganglionics hitch a ride on the internal carotid and then on its ophthalmic branch to the orbit, where they jump to the nasociliary nerve and finally leave the nasociliary as **long ciliary nerves** to the eye. Sympathetic impulses dilate the pupil and parasympathetics contract the pupil, but their individual roles in ciliary body function are unclear.

A small amount of smooth muscle is found in the superior tarsus and along the orbital floor. When contracted, the upper lid is pulled up and the eyeball pushed forward. This is a sympathetic effort to improve vision.

In the event of injury to the sympathetic optic input, the pupil becomes constricted due to the unopposed action of the parasympathetics. The upper lid droops and the eyeball recedes slightly (Horner's syndrome).

Nerves of the orbit. Three motor nerves have been noted: oculomotor, trochlear and abducens. All pass from the brain stem through the cavernous sinus and through the superior orbital fissure, from where they run to the extraocular muscles.

Sensory nerves from the ophthalmic division of the trigeminal (V1) enter the orbit. These include the frontal, lacrimal and nasociliary nerves. All three enter the orbit through the superior orbital fissure.

Frontal nerve. This 'continuation' of the ophthalmic passes forwards over the eye and divides into supraorbital and supratrochlear branches, which supply the upper lid, forehead and central scalp.

Nasociliary nerve. More deeply situated in the orbit, this nerve runs along the medial orbital border and pierces the ethmoid to reach the anterior nasal cavity. It finally emerges at the skin surface from under the nasal bone. It provides sensory fibers to the ethmoid, anterior nasal cavity mucosa and skin on the dorsum of the nose. At its entry into the orbit, it carries sympathetic fibers from the 'back' of the ophthalmic artery. These are passed to the eye as long ciliary nerves.

Lacrimal nerve. This small nerve runs along the lateral orbit after passing through the superior orbital fissure. After reaching the lacrimal gland, it passes to the skin of the upper lid. In its terminal portion, it receives postganglionic parasympathetic fibers from the zygomatic branch of V2 which it delivers to the lacrimal gland. The parasympathetic fibers come from the greater petrosal branch of the facial nerve via the pterygopalatine ganglion and zygomatic branch of V2.

Ophthalmic artery. This vessel branches from the internal carotid artery where the latter emerges from the cavernous sinus. It passes with the optic nerve through the optic foramen. The first branch of the ophthalmic is the central artery of the retina, which enters the optic nerve approximately 12 mm behind the eyeball. Multiple branches of the ophthalmic artery supply the extraocular muscles, ethmoid, anterior nasal cavity, and skin of the lids, nose, forehead and central scalp. In general, throughout its course, the ophthalmic artery branches accompany ophthalmic nerve branches.

Lacrimal apparatus. These are described in the Face section.

Ear. This consists of three areas: external ear and external auditory meatus; middle ear and ossicular chain; and inner ear containing neural balance (vestibular) and hearing (cochlear) organs.

All except the pinna of the external ear are housed in the temporal bone. The external ear collects sound waves and conveys them to the middle ear by means of the ear drum (tympanum) which stretches across the termination of the external auditory meatus. In the middle ear, three bony jointed lever arms (ossicles) mechanically transduce air waves into mechanical movements and transport them to the fluid-filled inner ear. Fluid waves in the inner ear stirred up by ossicular movements stimulate receptor hair cells (organ of Corti) in the cochlea which, in turn, send auditory signals to the brain via the cochlear portion of the statoacoustic nerve (VIII). This nerve travels in the internal auditory meatus in the petrous portion of the temporal bone. Balance and head movement sensors in the sacculus, utricle and semicircular canals of the inner ear signal the brain through the vestibular arm of the statoacoustic nerve.

Other features of the ear include the progress of the facial nerve (VII) through the petrous and mastoid portions of the temporal bone, and the connections of the middle ear with the nasopharynx and mastoid air cells.

External ear. The **pinna** (L. wing) or **auricle** is composed of skin with a cartilaginous skeleton. Important anatomical features include the outer cartilage-supported rim, the **helix**, which ends inferiorly in the soft **lobule**. Just anterior to and parallel with the helix is a second cartilage-supported ridge, the **antihelix**. In the center of the external ear is the opening of the **external auditory meatus**. Anterior to the meatal opening is the projecting cartilage-supported **tragus** and posterior to the meatus is the smaller **antitragus**. The lateral outer half of the external meatus is composed of skin and cartilage; the medial inner half is skin and bone. Ceruminous glands in the lateral part secrete the ear wax. The external meatus is around 2.5 cm in length in the adult. It terminates at the ear drum or **tympanum** (described with the middle ear). Vestigial auricular muscles are present in the scalp. These are of considerable use to some animals, but not to humans (except at parties).

Nerve supply to the external ear is offered by the auriculotemporal branch of the trigeminal, and greater auricular and lesser occipital branches of C2; interestingly, a part of the meatus is supplied by a branch of the vagus. Instrumental manipulation of the meatus in some people induces coughing or nausea. The vagus meatal signals are interpreted as originating in the larynx or stomach.

Blood supply to the external ear is obtained from the posterior auricular branch of the external carotid, the auricular branches of the superficial temporal artery, and a deep auricular branch of the maxillary.

Middle ear. This consists of an air-filled irregular box connected to the air-filled nasopharynx by the **auditory tube**, which enters the box anteriorly. Posteriorly is a bony opening or **aditus**, which leads to the **mastoid antrum** (Gk. cave). The antrum in turn communicates with the mastoid air cells which fill the mastoid process to a varying degree.

The **tympanic membrane** (ear drum or **tympanum**) separates the external meatus from the middle ear (**tympanic cavity**). The tympanic rim is set in the surrounding bone. When viewed through the external meatus by a lighted instrument (otoscope), the drum appears gray in color. The handle of the malleus is visible extending down its upper half. Extending anteroinferiorly from the malleus tip is a cone of reflected light which is obvious in the healthy drum. The middle ear is related anteroinferiorly to the internal carotid artery and inferiorly to the jugular bulb.

The **auditory ossicles** traverse the middle ear. These are the **malleus** (hammer), the **incus** (anvil) and the **stapes** (stirrup). The malleus handle is firmly attached to the drum, the head protruding upwards and jointed to the incus above the drum in a tiny synovial joint. The incus is anchored to bone by a short leg or crus, and is jointed to the head of the stapes by means of a longer crus and tiny synovial joint. The stapes footplate is firmly anchored to an opening in the medial wall of the tympanic cavity – the **oval window** or **fenestra vestibuli**. Two muscles are associated with the ossicular chain: **tensor tympani** enters the middle ear from a canal just above the auditory tube with its tendon inserted on the handle of the malleus; **stapedius** originates in the posterior wall of the tympanic cavity and inserts on the neck of the stapes. These muscles control the magnitude of movement of the ossicular chain.

Running across the top of the drum is the **chorda tympani** branch of the facial nerve bound for the infratemporal fossa (see below).

In the center of the medial wall is a rounded **promontory**, a bulge housing the first turn of the spiral cochlea of the inner ear. Under the lower side of the promontory is another bony window, the **fenestra cochleae** or **round window**. The round and oval windows effect the wave connection between the middle and inner ears (see below).

Now for some interesting and clarifying embryology. The external meatus is derived from the first pharyngeal cleft (gill slit). The middle ear and auditory tube are derived from the first pharyngeal pouch. The eardrum represents the persisting tissue between the cleft and pouch. The pharyngeal arches between pouches and cleft also contribute to the ear. The first arch contributes the malleus and incus, and tensor tympani muscle. The nerve of the first arch is the trigeminal (V). Thus, the tensor tympani is trigeminal-innervated. Second-arch remnants are the stapes and stapedius muscle. Thus, as the nerve of the second arch is the facial (VII), stapedius is supplied by the facial nerve. Note that the first and second arches 'come together' at the ear region, thereby accounting for the minute division between arch remnants and the facial and trigeminal nerves.

Facial nerve in the temporal bone. Cranial nerve VII travels along the internal auditory meatus opening in the posterior face of the petrous pyramid in company with cranial nerve VIII. In the bone above the promontory of the middle ear, the nerve produces its sensory **geniculate ganglion**. Just beyond the ganglion, the nerve turns sharply downwards and backwards, forming a 'knee' or **genu**. It then runs over the top of the middle ear and down behind the middle ear through the mastoid process in a bony **facial canal** to emerge through the stylomastoid foramen. Branches of the facial nerve in the temporal bone include the greater petrosal nerve, nerve to stapedius and the chorda tympani. The **greater petrosal nerve** containing parasympathetic preganglionic fibers issues from the geniculate ganglion (although it does not synapse with the neurons of the ganglion; the geniculate contains cell bodies of VII sensory branches only). The greater petrosal then passes through the petrous bone to emerge at the facial hiatus on its anterior face and, with the deep petrosal of the sympathetic system, forms the nerve of the pterygoid canal bound for the pterygopalatine ganglion. The **nerve of the stapedius** supplies that small muscle only. The **chorda tympani** branches off the facial trunk about half-way down the facial canal, retraces its way back to the tympanic cavity in a tiny canal parallel to the facial canal, and proceeds across the top of the eardrum and neck of malleus to enter a tiny bony canal in the anterior middle ear wall. The canal leads to the petrotympanic fissure in the mandibular fossa, from where progression is to the infratemporal fossa and lingual nerve.

The facial nerve is at risk during any surgical invasion of the mastoid process or middle ear, in middle ear–mastoid infections and in temporal bone fractures. Middle ear infection is usually caused by obstruction of the auditory tube due to nasopharyngeal infection and swelling – usually due to adenoidal tonsillar tissue. Swelling due to any cause around the nerve in the facial canal often produces temporary facial nerve paralysis.

Nerve and blood supply of the middle ear. The tympanic branch of the glossopharyngeal (IX) nerve does most of the work here, forming a **tympanic plexus** in the middle ear before leaving for the parotid gland. The external surface of the tympanum is largely supplied by the trigeminal. Blood is supplied largely by a tympanic branch of the maxillary artery, which enters through the petrotympanic fissure, and a stylohyoid branch of the posterior auricular artery, which enters through the stylohyoid foramen.

Inner ear. The organs of equilibrium and hearing are housed in a **bony labyrinth** in the petrous pyramid. The bony labyrinth consists of a central **vestibule**, which contains spaces for the **sacculus** and **utricle**. The oval window of the middle ear opens into the vestibule. Posterosuperiorly are three **semicircular canals**, each at right angles to the others. Anteromedially is the **cochlea** (L. snail shell). Around a central bony **modiolus** are channels for a spiral tube that makes two and three-quarter turns upwards to an apex. The basal turn of this cochlear spiral projects into the tympanic cavity as the promontory. The bony labyrinth is filled with thin fluid **perilymph**, thought to be derived from cerebrospinal fluid through a thin connecting membranous sac in the neighboring dura. Floating in the perilymph of the osseous labyrinth is the **membranous labyrinth**, which is filled with a chemically different fluid, the **endolymph**. The membranous labyrinth contains the organs of balance and hearing. The three membranous semicircular canals contain hair cells in swellings in each canal (**ampullae**). These are sensitive to head movements. Because of fluid inertia, the endolymph in the semicircular canal presses against hair cells upon movement of the head. Changes in hair cell position stimulate nerve impulses, which report to the brain via the vestibular portion of nerve VIII. Cell bodies of the hair cells are located in a **vestibular ganglion** in the ampullae.

The middle part of the membranous labyrinth, housed in the bony vestibule, consists of two small sacs, the **utricle** and **sacculus**. These organs contain gravity-affected grains of calcium carbonate – **otoliths**. Otoliths stimulate hair cells in these structures when they are stressed by gravity upon changes of head position, and with acceleration and deceleration of the head. The cell bodies of these hair cells are in the walls of the sacculus and utricle. They, too, contribute to nerve output in the vestibular part of VIII.

Finally, there is the cochlear portion of the membranous labyrinth. The cochlear membranous tube, filled with endolymph, is the **cochlear duct** or **scala media**. It is fixed to the central modiolus and opposite bony wall (**spiral ligament**) so that it divides the surrounding perilymph-filled channel into two parts, the superior **scala vestibuli** and the inferior **scala tympani**. A cluster of capillaries, the **stria vascularis**, in the spiral ligament may be the source of the endolymph of the membranous labyrinth. The boundaries of the cochlear duct are an upper **vestibular membrane** (**Reissner's membrane**) between the scala vestibuli and cochlear duct, and a **basilar membrane** between the cochlear duct and scala tympani. The two perilymph-filled scala are connected at the apex of the two and three-quarter cochlear turns about the modiolus. The connection is termed the **helicotrema** (*helico* = Gk. spiral and *trema* = hole). Fluid waves entering the scala vestibuli can travel up to the helicotrema and back down the scala tympani. The moving stapes in the oval window induces fluid waves in the perilymph to follow such a path. The scala tympani ends at the membrane-covered **round window**. Sound waves are stilled by movement of the round window membrane. Thus, there is no rebound of these waves.

The hearing organ, the **organ of Corti**, is housed in the cochlear duct (scala media). It rests on the basement membrane. Fluid wave-induced movement of the vestibular or basilar membranes stimulates receptors in the organ of Corti. Different areas of the membranes respond to different fluid wavelengths. Thus, different areas respond to different tone pitches. The cell bodies of the receptor neurons of the organ of Corti lie in the **spiral ganglion** at the modiolar side of the organ. Central processes travel down canals in the modiolus and gather at the lateral opening of the internal auditory meatus to form the cochlear portion of nerve VIII. Thus, there are two fluid-filled labyrinths in the inner ear. In each labyrinth, all parts are connected and filled by perilymph or endolymph. There is, however, no fluid connection between the two labyrinths. The bony labyrinth and its perilymph are connected to the middle ear ossicular chain through the oval window.

Deafness may be due to obstruction of the external meatus (wax, foreign bodies), scarring of the tympanum or arthritic degenerative processes involving the interossicular joints. This type of hearing defect is termed **conduction deafness**. Disease or disorder of the organ of Corti or nerve VIII itself produces **nerve deafness**. Conduction of sound waves to the perilymph occurs through the bones of the calvaria (**bone conduction**) and supports the usual air conduction. Conduction deafness may be improved by removal of foreign bodies, use of hearing aids or surgery. Nerve deafness is not reversible.

Review Exercises

Figures 8.1–8.10 should be carefully studied and memorized, and then reproduced.

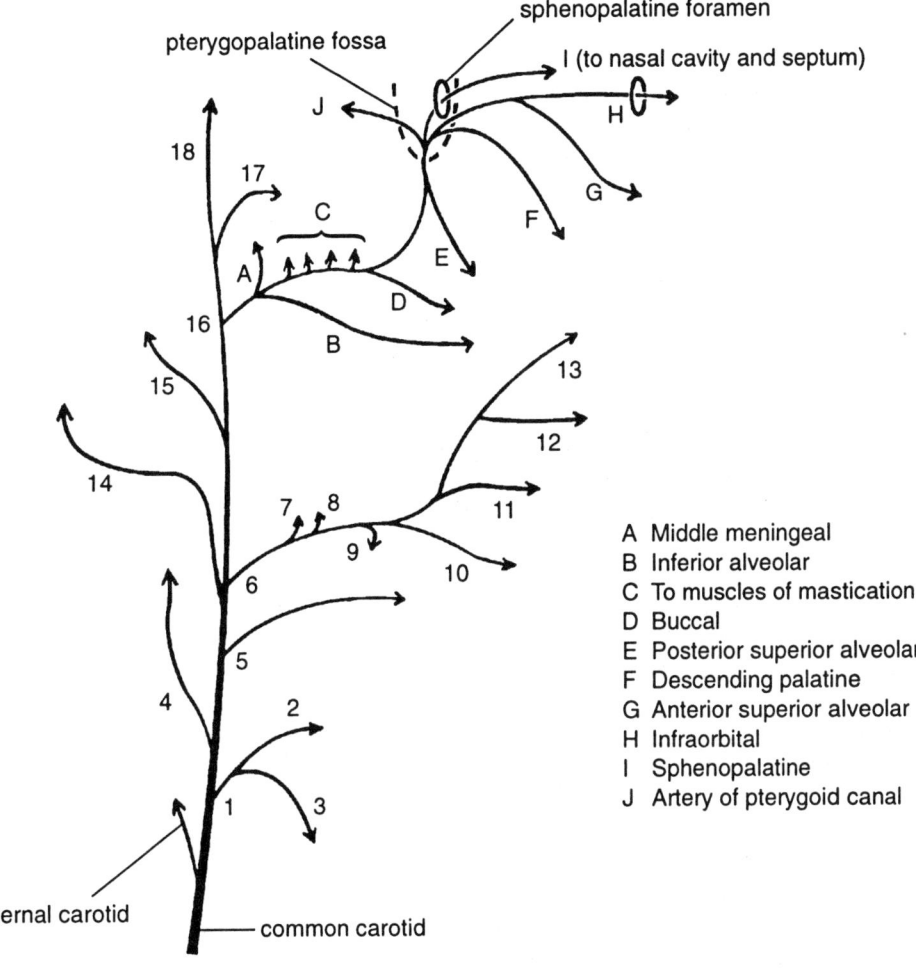

A Middle meningeal
B Inferior alveolar
C To muscles of mastication
D Buccal
E Posterior superior alveolar
F Descending palatine
G Anterior superior alveolar
H Infraorbital
I Sphenopalatine
J Artery of pterygoid canal

1. Superior thyroid
2. Superior laryngeal branch
3. Thyroid branch
4. Ascending pharyngeal
5. Lingual
6. Facial
7. Ascending palatine
8. Tonsillar
9. To submandibular gland
10. Submental
11. Inferior labial
12. Superior labial
13. Angular
14. Occipital
15. Posterior auricular
16. Maxillary
17. Transverse facial
18. Superficial temporal

Figure 8.1 External carotid artery (numbers) and maxillary artery (letters)

233

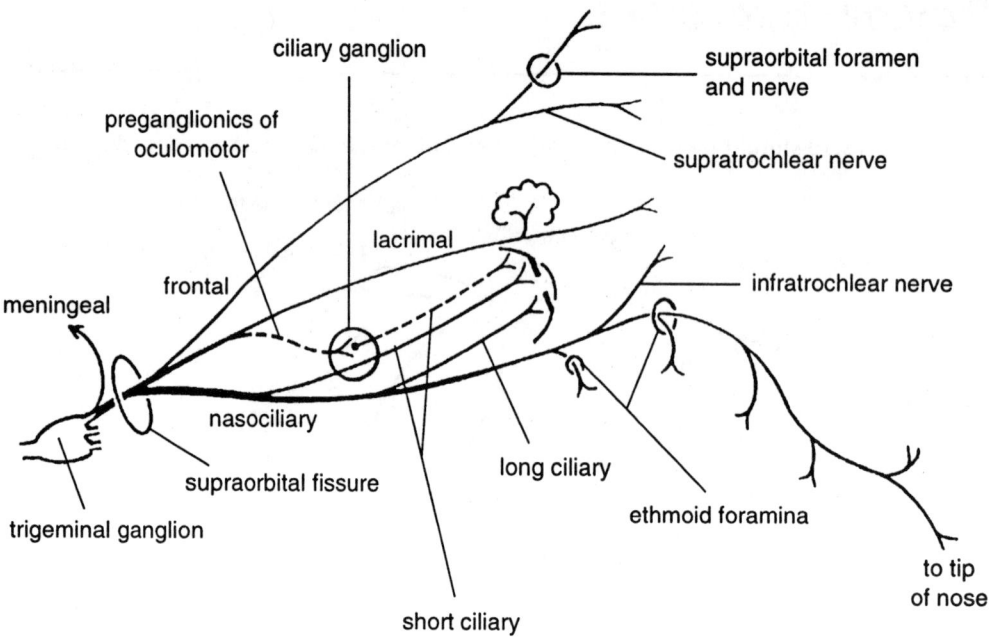

Figure 8.2 Trigeminal nerve, ophthalmic division V1

1. Purely sensory nerve
2. Enters orbit through the superior oribital fissure
3. Meningeal branch from nerve trunk
4. **Frontal nerve:** Supraorbital branch through supraorbital foramen to skin of forehead and scalp; supratrochlear branch to lids, conjunctiva and nose base
5. **Lacrimal nerve:** Sensory to upper lids; carries postganglionic parasympathetic fibers from pterygopalatine ganglion to lacrimal gland as a favor to the facial nerve; parasympathetic fibers reach lacrimal nerve via zygomatic branch V2
6. **Nasociliary nerve:** Sensory to eye structures via the long and short ciliaries; the latter run through the ciliary ganglion without synapse.

    In addition to sensory fibers, the short ciliaries carry postganglionic parasympathetics from the ciliary ganglion to the ciliary body and iris. Preganglionics are delivered to the ciliary ganglion by the oculomotor nerve.

    The long ciliary, which bypasses the ciliary ganglion, carries, in addition to its sensory fibers, postganglionic sympathetic fibers for dilator pupillae from the superior cervical ganglion. The infratrochlear branch innervates lids and conjunctivae, and the sides of the nose. Posterior and anterior ethmoids pass through ethmoid foramina to innervate structures of the nasal cavity. The nerve continues down the anterior nasal cavity, emitting internal nasal sensory branches, and ends at the skin of the dorsum and tip of the nose as external nasal branches

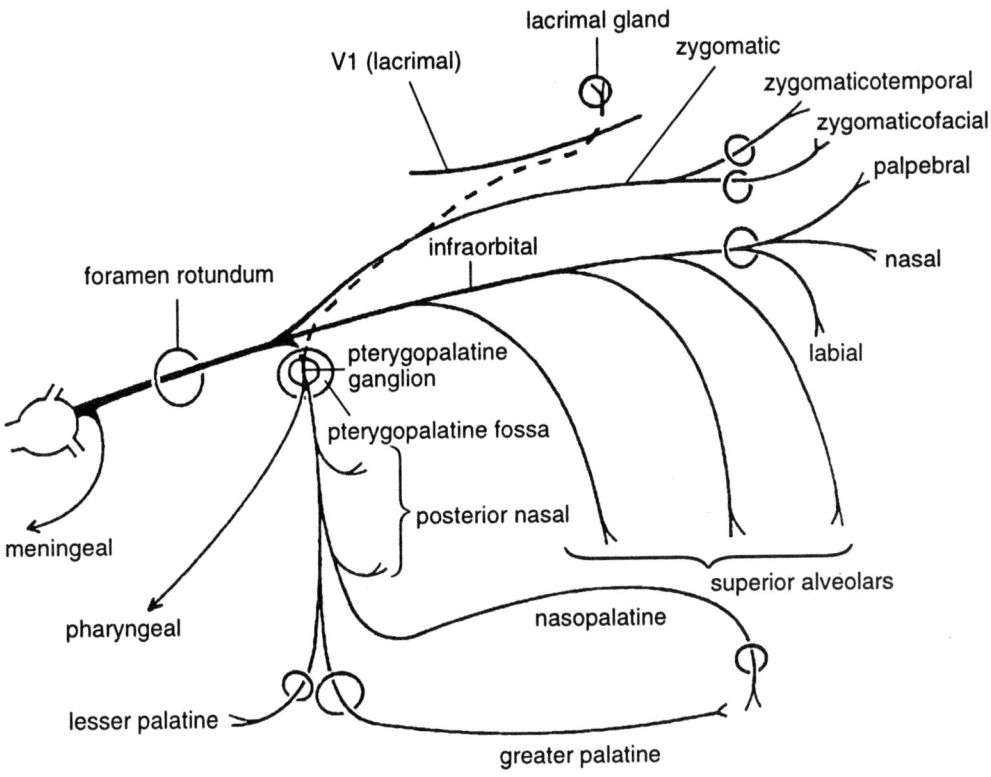

Figure 8.3 Trigeminal nerve, maxillary division V2

1. Purely sensory nerve
2. Enters pterygopalatine fossa through foramen rotundum
3. **Infraorbital nerve** passes through infraorbital fissure and canal before attaining skin via infraorbital foramen
4. **Zygomatic nerve** exits in two branches through zygomaticofacial and zygomaticotemporal foramina to supply skin over cheek and temple (note dashed line indicating communication carrying postganglionic parasympathetics from pterygopalatine ganglion to lacrimal gland)
5. Pterygopalatine ganglion is indicated in pterygopalatine fossa. V2 branches pass through the ganglion without synapse, then branch to nasal cavity, teeth and palate. **Nasopalatine nerve** passes through the incisive canal to reach the hard palate. **Greater** and **lesser palatine nerves** pass through greater and lesser palatine foramina to reach the hard and soft palate, respectively
6. Pharyngeal nerve innervates the nasopharyngeal mucosa

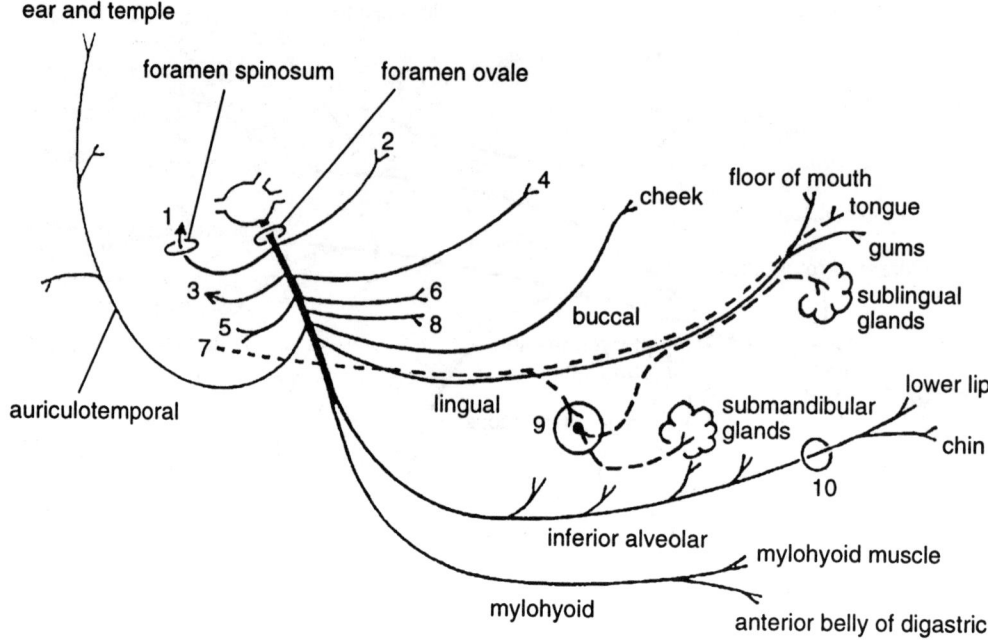

Figure 8.4 Trigeminal nerve, mandibular division V3 (mixed sensory-motor nerve; enters infratemporal fossa via foramen ovale)

1. Meningeal branch through foramen spinosum with middle meningeal artery
2. Deep temporal nerve to temporalis muscle
3. Nerve to tensor tympani muscle
4. Nerve to tensor veli palatini
5. Masseteric nerve
6. and 8. Medial and lateral nerves to pterygoid muscles
7. Chorda tympani from facial trunk; joins lingual in infratemporal fossa en route to submandibular ganglion, submandibular and sublingual salivaries, and taste buds on anterior two-thirds of tongue. Lingual proper is general somatic afferent (GSA) to anterior two-thirds of tongue and floor of mouth.
9. Submandibular ganglion
10. Inferior alveolar nerve passes through mental foramen; becomes mental nerve to lower lip and chin

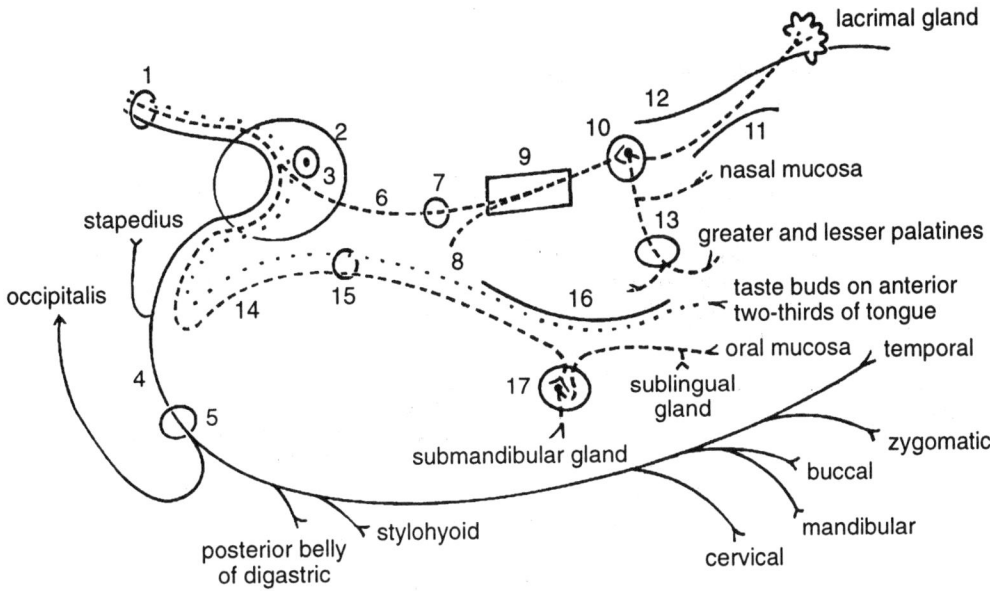

Figure 8.5   Facial nerve (VII) (SVE, SVA, GVE)

1. Internal auditory meatus
2. Middle ear
3. Geniculate ganglion
4. Facial canal in mastoid process
5. Stylomastoid foramen
6. Greater petrosal nerve
7. Facial hiatus in petrous pyramid
8. Deep petrosal nerve (sympathetic postganglionics)
9. Pterygoid canal
10. Pterygopalatine fossa and ganglion
11. Zygomatic nerve of V2 } carry postganglionics from pterygopalatine ganglion
12. Lacrimal nerve of V1 } to lacrimal gland
13. Greater and lesser palatine foramina
14. Chorda tympani (sensory cell bodies are in geniculate ganglion)
15. Petrotympanic fissure
16. Lingual nerve
17. Submandibular ganglion

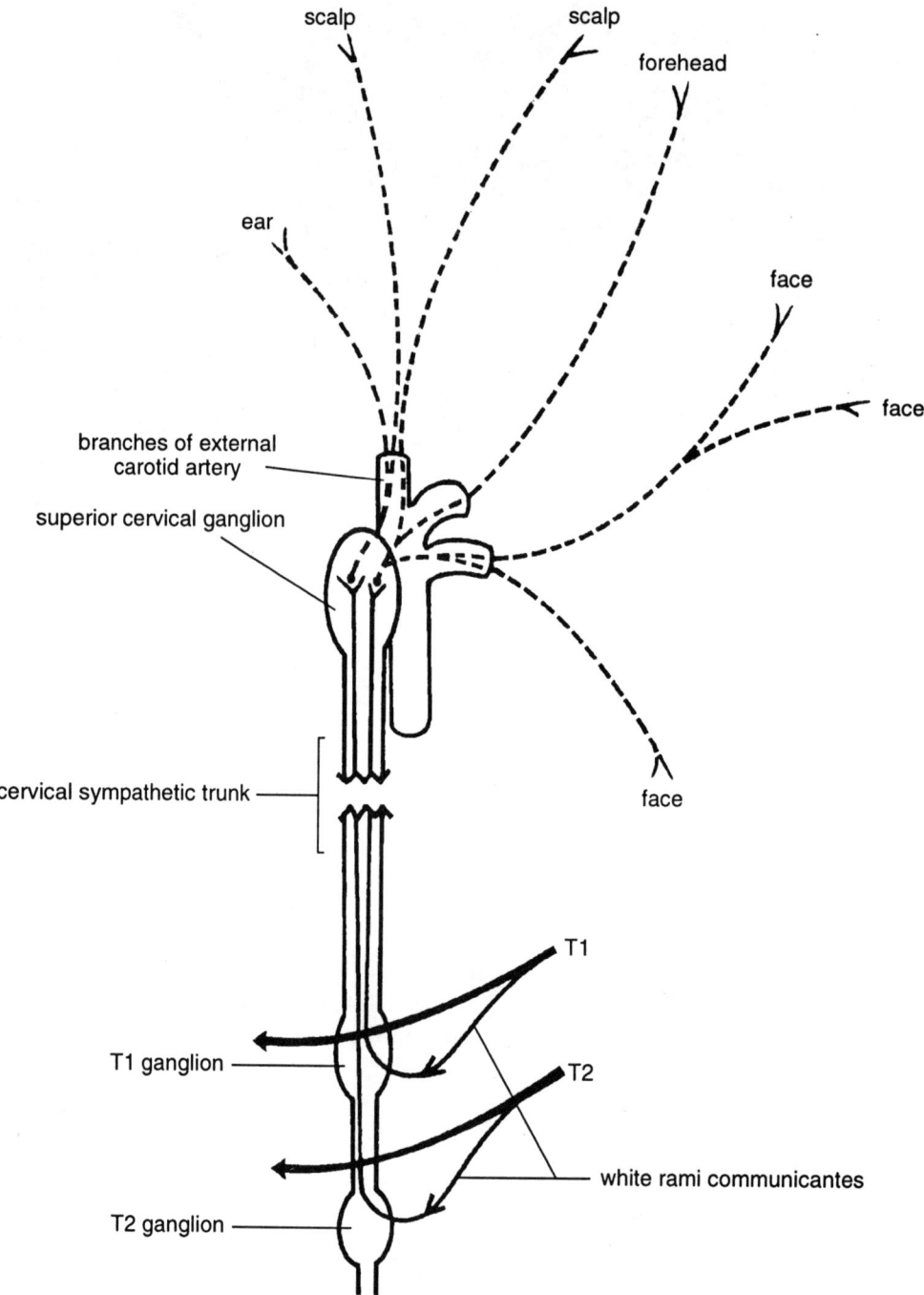

Figure 8.6 Sympathetics to head for 'peripheral' vasomotor, sudomotor and pilarrector function

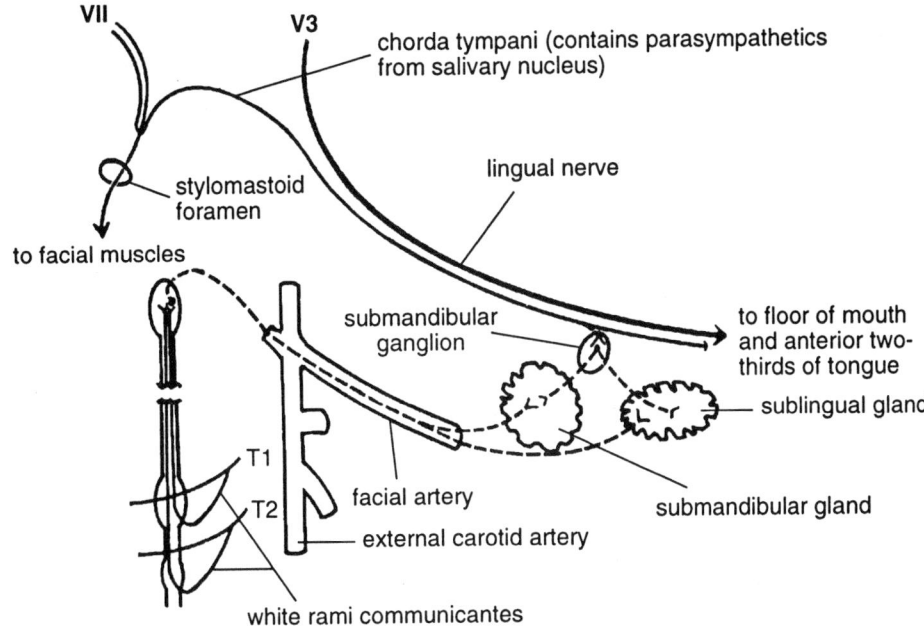

Figure 8.7 Autonomics to submandibular and sublingual salivary glands (submandibular ganglion)

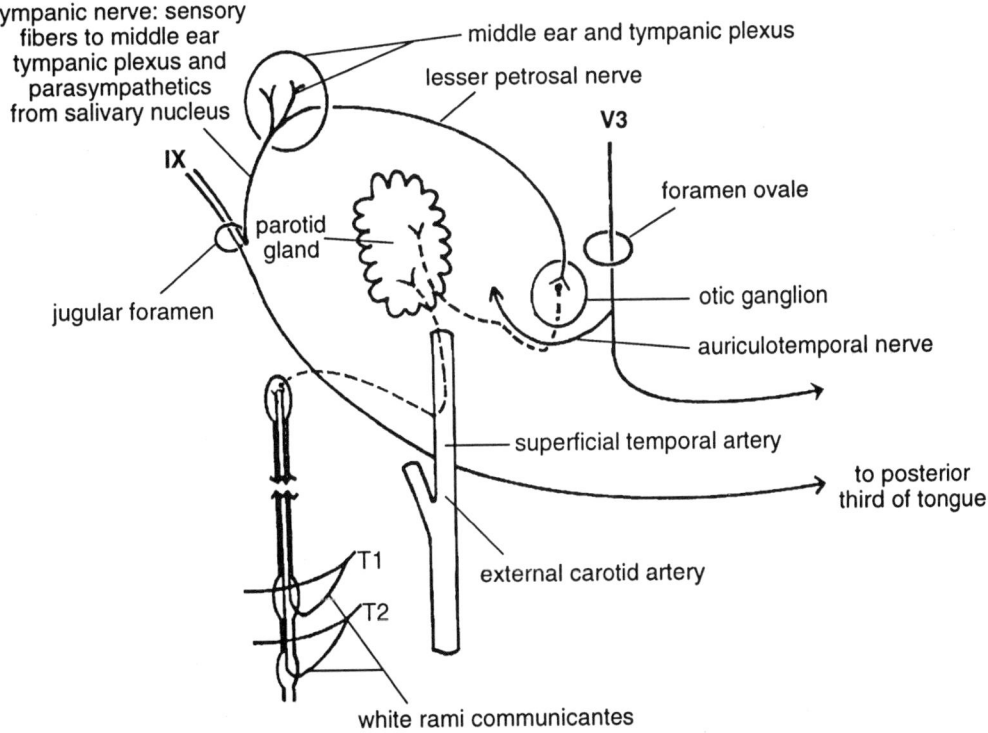

Figure 8.8 Autonomics to parotid gland (otic ganglion)

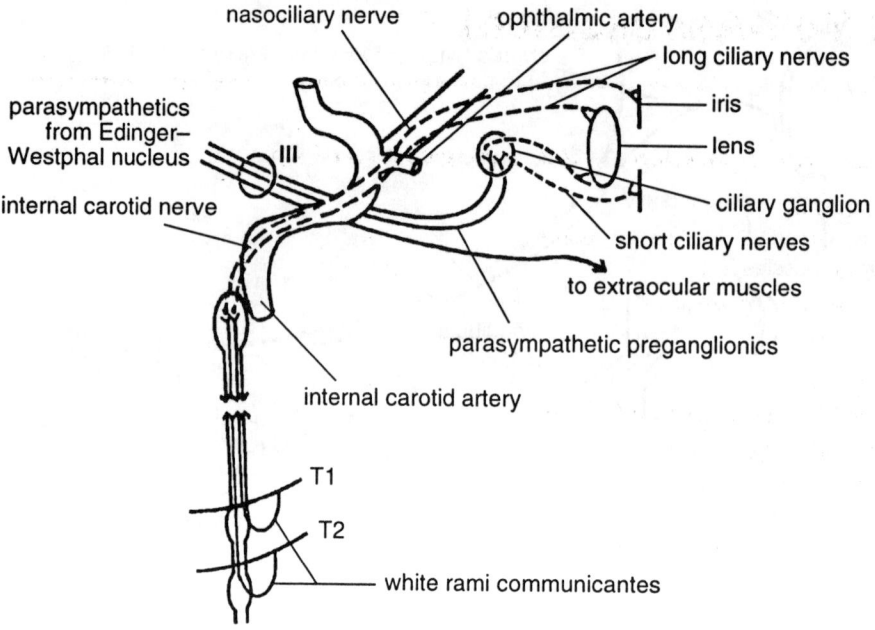

Figure 8.9 Autonomics to lens (ciliary body) and iris smooth muscle (ciliary ganglion)

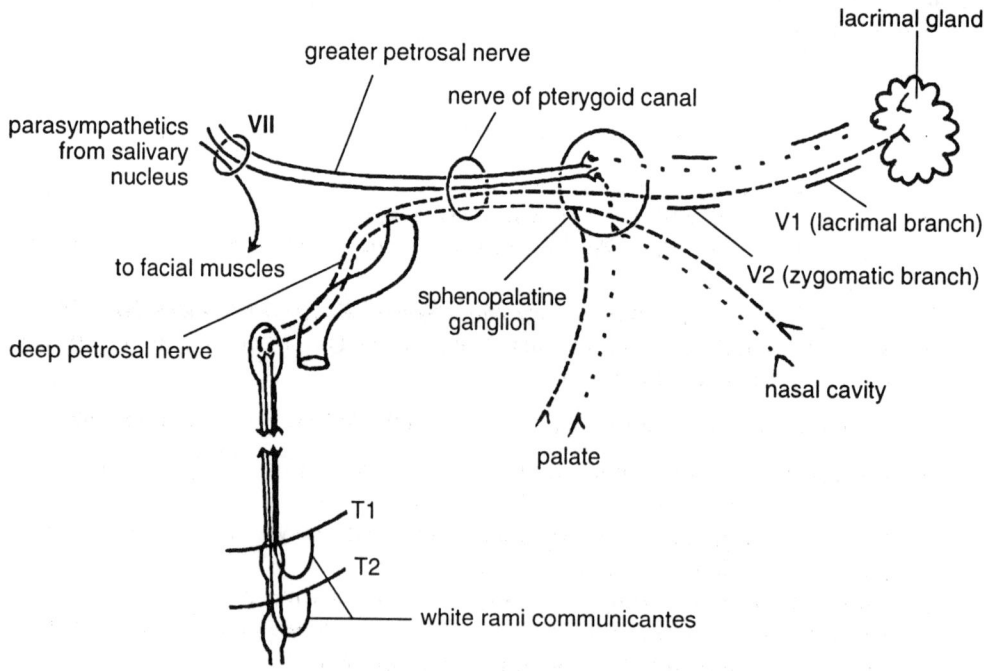

Figure 8.10 Autonomics to mucosal and lacrimal glands (sphenopalatine ganglion)

Head Self-Assessment

For each numbered item, select the closest related lettered item.

A   Scalp skin
B   Subcutaneous connective tissue
C   Galea aponeurotica
D   Loose areolar tissue
E   Pericranium

1. Gaping wound.
2. Localized but widespread infection.
3. Location of blood vessels.
4. Frontalis muscle.

A   Anterior cranial fossa
B   Posterior cranial fossa
C   Petrous pyramid
D   Orbit
E   Middle cranial fossa

5. Temporal lobes.
6. Cerebellum.
7. Cribriform plate.
8. Cavernous sinus.

9. Regarding the bones of the orbit, one of the following statements is incorrect:

A   The maxilla and zygoma form the inferior orbital rim.
B   The strongest part of the orbit is its medial wall.
C   X-ray evidence of blood in the maxillary sinus is a frequent finding in orbital floor fracture.
D   The optic foramen and superior orbital fissure occupy the apex of the orbit.
E   The infraorbital fissure, groove, canal and foramen afford passage for the terminal branch of the maxillary nerve.

10. Regarding the temporomandibular joint, one of the following statements is incorrect:

A   Is supported by the sphenomandibular and stylomandibular ligaments.
B   Is opened by action of the lateral pterygoid muscle.
C   Contains an intra-articular disc which serves to restrict the range of motion of the jaw, thus guarding against dislocation.
D   The posterior half of the mandibular fossa is non-articular.
E   Dislocation occurs when the condyle of the mandible passes forwards over the tubercle of the zygomatic process of the temporal bone.

11. The facial nerve trunk was inadvertently divided during an operation to remove the parotid gland. One of the following will occur on the side of the injury:

   A Loss of taste over the anterior two-thirds of the tongue.
   B A dry eye due to loss of lacrimal gland secretions.
   C Spastically closed eyelids.
   D Dry mouth due to operative loss of parotid, plus paralysis of submandibular and sublingual gland function.
   E Drooling.

12. Regarding the trigeminal nerve:

   A Innervates face and scalp back to the level of the ears.
   B Passes through the cavernous sinus before dividing into its three terminal branches.
   C Innervates the buccinator muscle through its buccinator branch.
   D Is responsible for general sensation in the entire mouth and oropharynx.
   E Contains motor fibers only for muscles of mastication.

13. Regarding autonomic fibers of the facial nerve:

   A Cell bodies for these fibers lie in the geniculate ganglion.
   B Preganglionic fibers travel with the internal carotid artery.
   C Preganglionic fibers are found in both the chorda tympani and the greater petrosal nerve.
   D Synapse is in the ciliary ganglion for postganglionic fibers providing para-sympathetic eye functions.
   E Are responsible for stimulating the secretion of all salivary glands.

14. Regarding the maxillary artery, choose the one incorrect statement:

   A Furnishes the bulk of blood to the submandibular gland.
   B Traverses the infratemporal fossa, pterygomaxillary fissure and sphenopalatine foramen.
   C Nourishes the calvaria and dura with blood.
   D As the sphenopalatine artery, it nourishes the structures of the nasal cavity.
   E Is responsible for blood supply of the upper and lower teeth.

15. Division of the right lingual nerve will produce:

   A Protrusion of the tongue to the right.
   B Loss of taste over the posterior third of the tongue.
   C Diminution of salivary secretion from the opening of the duct opposite the upper second molar.
   D Loss of pain and temperature sensation over the right side of the tongue.
   E Deviation of the uvula to the right on testing the gag reflex.

16. In the process of deglutition, choose the one incorrect statement:

    A Muscles of the palatoglossal and palatopharyngeal arches are important in preventing oral regurgitation.
    B Suprahyoid muscles elevate and fix the hyoid, followed by elevation of the larynx by thyrohyoid, palatopharyngeus and stylopharyngeus.
    C Closure of the laryngeal airway is accomplished by elevation of the larynx, depression of the epiglottis, and contraction of aryepiglottic and intrinsic laryngeal muscles.
    D Injury to the recurrent laryngeal nerve will have no effect on the smooth performance of deglutition.
    E The pharyngeal constrictors, innervated by the pharyngeal plexus, provide the power to move a bolus of food into the esophagus.

17. When an eye is adducted, which one of the following is capable of depressing it?

    A Superior rectus muscle.
    B Lateral rectus muscle.
    C Medial rectus muscle.
    D Superior oblique muscle.
    E Inferior oblique muscle.

18. Destruction of the oculomotor nerve may cause all of the following except:

    A Xerophthalmia (drying of the cornea).
    B Dilation of the pupil.
    C Ptosis.
    D Lateral deviation of the eye.
    E Loss of accomodation.

19. Regarding the sympathetic innervation of the eye, one of the following statements is incorrect:

    A Preganglionic neurons are located in the superior cervical ganglion.
    B Sympathetic fibers reach the eye by riding on the internal carotid and ophthalmic arteries.
    C Sympathetic fibers may pass through the ciliary ganglion.
    D Destruction of the stellate ganglion results in constriction of the pupil.
    E Postganglionic sympathetic fibers are found in both long and short ciliary nerves.

20. Regarding the middle ear, one of the following statements is incorrect:

    A Mastoid air cells, the middle ear and the nasopharynx are connected by means of the aditus ad antrum and the eustachian tube.
    B The middle ear receives sensory innervation from the glossopharyngeal nerve.
    C Middle ear ossicles transmit motion of the tympanum to the perilymph of the inner ear via the footplate of the stapes and the oval window.
    D The tensor tympani and stapedius muscles can control the amplitude of ossicular motion.
    E The geniculate ganglion supplies sensory cell bodies for general sensation over the anterior two-thirds of the tongue.

21. Regarding the meninges and cerebrospinal fluid, one of the following statements is incorrect:

    A   The dura is as loosely applied to the surrounding bone of the calvaria as they are to the vertebral bodies.

    B   The dural falx cerebri extends between the cerebral hemispheres.

    C   Dural venous sinuses drain brain blood and cerebrospinal fluid to the internal jugular vein.

    D   The cerebrospinal fluid may be safely sampled by needle aspiration in the subarachnoid space between the conus medullaris and S2.

    E   The cerebrospinal fluid is formed within the brain, but escapes to bathe and float the entire central nervous system.

22. Regarding the blood supply of the brain, one of the following statements is incorrect:

    A   Consists of arterial blood from vertebral and internal carotid arteries.

    B   The carotid canal traverses the petrous portion of the temporal bone.

    C   The internal carotid artery is contained within the cavernous sinus along with cranial nerves III, IV, VI, V1 and V2.

    D   Vertebral artery branches include the spinal, cerebellar and posterior cerebral arteries.

    E   The vertebral and internal carotid systems communicate through the circle of Willis.

Match the numbered items with closely related lettered items. Lettered items may be used more than once or not at all.

    A   Third upper molar tooth
    B   Maxillary sinus
    C   Mandibular division of trigeminal
    D   Lacrimal gland
    E   Sella turcica
    F   Incisive canal
    G   Nerve of the pterygoid canal

23. Greater palatine foramen.

24. Nasopalatine nerve.

25. Ophthalmic division of trigeminal.

26. Nasal middle meatus.

27. Sphenopalatine ganglion.

28. Sphenoid sinus.

Match numbered items with closely related lettered items. Lettered items may be used more than once or not at all.

A   Injury to the facial nerve at the internal auditory meatus
B   Injury to maxillary nerve at the foramen rotundum
C   Injury to intracranial glossopharyngeal nerve
D   Injury to trigeminal nerve between brain and trigeminal ganglion
E   Injury to facial nerve at the stylomastoid foramen
F   Injury to mandibular nerve between trigeminal ganglion and foramen ovale
G   Injury to the glossopharyngeal nerve 1 cm inferior to the jugular foramen

29. Paralyzed facial muscles, loss of lacrimal secretion.

30. Paralyzed facial muscles, no loss of taste on the anterior two-thirds of tongue.

31. Loss of general sensation in the middle ear, loss of parotid secretion.

32. Lateral deviation of the uvula, anesthesia of the lower eyelid.

33. Anesthesia of the posterior oropharynx, no loss of parotid secretion.

34. Anesthesia of the anterior two-thirds of the tongue, normal sensation on the tip of the nose.

Match numbered items with closely related lettered items. Lettered items may be used more than once or not at all.

A   Foramen spinosum
B   Pterygomaxillary fissure
C   Stylomastoid foramen
D   Foramen rotundum
E   Foramen lacerum
F   None of the above

35. Mandibular nerve

36. Facial nerve

37. Maxillary nerve

38. Chorda tympani

39. Middle meningeal artery

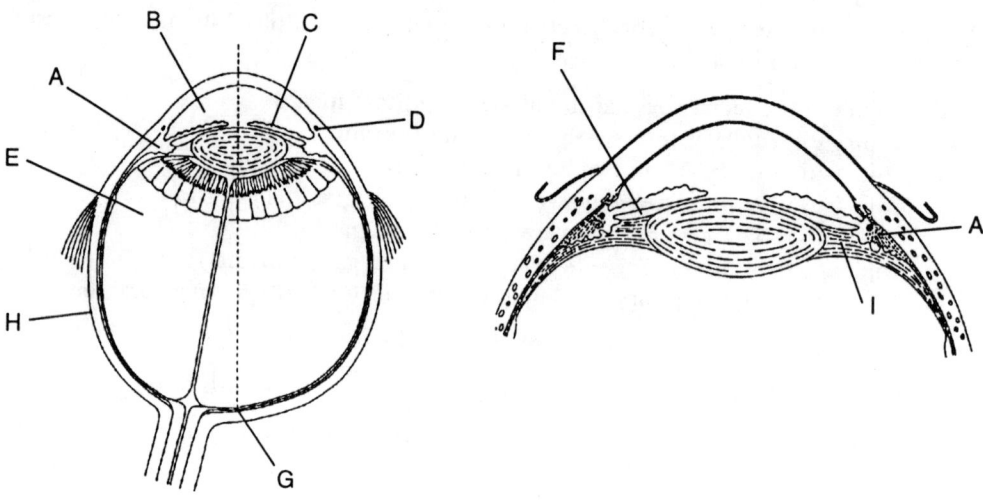

Identify on the above diagrams:

40. Iris

41. Ciliary body

42. Sclera

43. Anterior chamber

44. Posterior chamber

45. Vitreous body

46. Macula

47. Suspensory ligaments of lens

Identify on diagram:

48. Malleus

49. Oval window

50. Eustachian tube

51. Cochlea

52. Tensor tympani

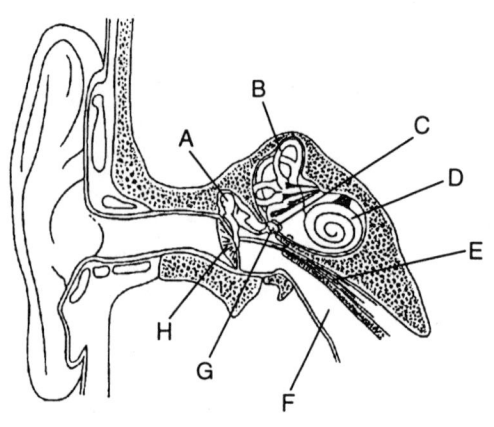

Answers and Explanations

1. **C**

2. **D** Infection may spread over the entire scalp with initial involvement of this space at any point.

3. **B** The rich blood supply to the scalp is carried by vessels in this layer.

4. **C**

5. **E**

6. **B**

7. **A**

8. **E**

9. **B** The medial orbital wall is largely composed of paper-thin labyrinthine elements of the ethmoid bone. The inferior wall (floor) constitutes the roof of the maxillary sinus. It, too, is thin and, if injured, may well bleed into the maxillary sinus.

10. **C** The disc affords an 'extra-articular' area for increased range of motion. The posterior part of the mandibular fossa is non-articular. It contains the petrotympanic fissure for passage of the chorda tympani to the infratemporal fossa.

11. **E** Loss of function of buccinator and perioral muscles causes a sagging opening at the corner of the mouth and drooling of saliva. The chorda tympani (taste and submandibular/sublingual secretion) and the greater petrosal nerve (lacrimation) are given off before the area of truncal injury described. With continued secretion of the submandibular and sublingual glands, there will be no noted dryness of the mouth. Paralysis of the ipsilateral orbicularis oculi leaves the eye open.

12. **A** Occipital and postauricular nerves handle the scalp posterior to the ears (C2–3). The trigeminal divides into its three terminal branches before the first and second branches (ophthalmic and maxillary) enter the cavernous sinus. The mandibular division escapes the sinus entirely. Buccinator receives innervation from the facial nerve. The trigeminal oral sensory responsibility ceases in the oropharynx where the glossopharyngeal takes over. Trigeminal innervates the tensors tympani and veli palatini, mylohyoid and anterior belly of digastric as well as the muscles of mastication.

13. **C** The geniculate neurons are all sensory (chorda tympani). The cell bodies for facial nerve preganglionic fibers are in the brain stem. They do not use internal carotid transportation help. The ciliary ganglion provides postganglionic neurons for fibers in the oculomotor nerve. Glossopharyngeal nerve stimulates secretion of the parotid.

14. **A** The submandibular gland is primarily supported by the facial artery. The middle meningeal branch of the maxillary, passing through the foramen spinosum, supplies the dura and calvaria.

15. **D** The lingual nerve does not influence tongue protrusion. The trigeminal nerve to the tensor of the soft palate is given off from the mandibular division trunk before the latter separates into its inferior alveolar and lingual branches. Even if the soft palate tensor were paralyzed, the deviation would be to the left (unopposed pull of the intact tensor). Taste over the posterior third of the tongue is provided by the glossopharyngeal nerve. The latter also innervates the parotid gland, the duct of which enters the mouth opposite the second upper molar tooth.

16. **D** Loss of recurrent laryngeal nerve function causes paralysis of the ipsilateral intrinsic laryngeal muscles with incomplete closure of the laryngeal airway. Aspiration of swallowed food occurs. Muscles of the faucial arches close the fauces and prevent regurgitation of food back into the mouth.

17. **D** With adduction, the axis of the eye and axis of pull of the superior oblique are in line. Thus, pull of the superior oblique rolls the adducted eye downwards.

18. **A** Lacrimation is a facial nerve function. Accomodation requires convergence, pupillary constriction and rounding of the lens, all caused by smooth muscle innervated by the oculomotor nerve. Unopposed pull of dilator pupillae (sympathetic system) and lateral rectus (abducens nerve) produces dilation and lateral deviation. Ptosis results from paralysis of the oculomotor-innervated levator palpebri muscle.

19. **A** Preganglionic sympathetic neurons are located in lateral horn cells of T1–2 spinal cord levels. The superior cervical sympathetic ganglion supplies postganglionic neurons for all sympathetics entering the head. The postganglionic fibers may pass through the ciliary ganglion, but they do not synapse there. Sympathetic postganglionic fibers reach the eye via both long and short ciliary nerves.

20. **E** General sensation of the anterior two-thirds of the tongue is supplied by the trigeminal lingual branch. The geniculate ganglion, at the inner ear genu of the facial nerve, contains cell bodies for the sensory functions of the facial nerve (principally taste over the anterior two-thirds of the tongue).

21. **A** The dura is tightly applied to the calvaria, but loosely applied to the vertebral bony borders of the spinal canal. Cerebrospinal fluid gains access to venous blood of the dural sinuses through the arachnoid villi. There is no spinal cord below the conus medullaris. The lumbar CSF cistern extends to S2. Sampling here will harmlessly tweak only fibers of the cauda equinus.

22. **D** The posterior cerebral artery is a branch of the basilar artery, the latter formed by union of the right and left vertebral arteries. The internal carotid has four portions: cervical, petrous, cavernous and cerebral. The cervical portion has no branches. The petrous portion traverses the petrous portion of the temporal bone in the bony carotid canal. The cavernous part is contained in the cavernous sinus. Communicating arteries join vertebral (basilar) and carotid systems (posterior communicators), and right and left carotid systems (anterior communicator) to form the circle of Willis.

23. **A** Foramen for passage of the greater palatine nerve is located in the hard palate here.

24. **F** For the anterior palatine branches of the nerve.

25. **D** The V1 lacrimal nerve carries postganglionic parasympathetic fibers from the pterygopalatine ganglion to the lacrimal gland as a favor to the facial nerve.

26. **B** Site of drainage of the maxillary sinus.

27. **G** Nerve of the pterygoid canal carries facial preganglionics to the pterygopalatine ganglion. (Could be D as postganglionics from the ganglion innervate the lacrimal gland.)

28. **E** The sphenoid sinus is immediately anterior to the sella and its contained pituitary gland.

29. **A** Injury would have to be medial to the middle ear take-off of the greater petrosal nerve to influence lacrimal secretion.

30. **E** The facial nerve must be injured distal to the take-off of the chorda tympani (in the facial canal in the mastoid process of the temporal bone) to paralyze facial muscles, but preserve chorda tympani taste function.

31. **C** Glossopharyngeal sensory fibers to the mucosa of the middle ear and preganglionic autonomic fibers to the parotid (all carried in the tympanic branch of IX) are given off in the jugular foramen. To knock out the tympanic nerve, injury must be proximal to the jugular foramen.

32. **D** Lower eyelid is innervated by the trigeminal maxillary division infraorbital nerve. Tensor palatini innervation is given off from the trigeminal mandibular nerve trunk just after passage through the foramen ovale. Thus, V2 and V3 must be destroyed in the injury.

33. **G** The glossopharyngeal nerve supplies both parotid and oropharynx. As the parotid innervation contained in the tympanic nerve is given off in the jugular foramen, a lesion 1 cm below the foramen would spare the tympanic nerve and affect only the oropharynx.

34. **F** The lingual nerve from V3 mandibular nerve is destroyed, resulting in anesthesia of the anterior two-thirds of the tongue. The tip of the nose is supplied by the V1 nasociliary nerve which remains intact.

35. **F** Uses the foramen ovale.

36. **C**

37. **D**

38. **F** Passes through the petrotympanic fissure of the posterior mandibular fossa.

39. **A**

40. **C**

41. **A**

42. **H**

43. **B**

44. **F**

45. **E**
46. **G**
47. **I**
48. **A**
49. **C**
50. **F**
51. **D**
52. **E**